BURLEIGH DODDS SCIENCE: INSTANT INSIGHTS

NUMBER 56

Sustainable forest management

burleigh dodds
SCIENCE PUBLISHING

Published by Burleigh Dodds Science Publishing Limited
82 High Street, Sawston, Cambridge CB22 3HJ, UK
www.bdspublishing.com

Burleigh Dodds Science Publishing, 1518 Walnut Street, Suite 900, Philadelphia, PA 19102-3406, USA

First published 2022 by Burleigh Dodds Science Publishing Limited
© Burleigh Dodds Science Publishing, 2022. All rights reserved.

Notice
No responsibility is assumed by the publisher for any injury and/or damage to persons or property as a matter of product liability, negligence or otherwise, or from any use or operation of any methods, products, instructions or ideas contained in the material herein.

British Library Cataloguing in Publication Data
A catalogue record for this book is available from the British Library

ISBN 978-1-80146-408-6 (Print)
ISBN 978-1-80146-409-3 (ePub)

DOI: 10.19103/9781801464093

Typeset by Deanta Global Publishing Services, Dublin, Ireland

Contents

Series list

Title	Series number
Sweetpotato	01
Fusarium in cereals	02
Vertical farming in horticulture	03
Nutraceuticals in fruit and vegetables	04
Climate change, insect pests and invasive species	05
Metabolic disorders in dairy cattle	06
Mastitis in dairy cattle	07
Heat stress in dairy cattle	08
African swine fever	09
Pesticide residues in agriculture	10
Fruit losses and waste	11
Improving crop nutrient use efficiency	12
Antibiotics in poultry production	13
Bone health in poultry	14
Feather-pecking in poultry	15
Environmental impact of livestock production	16
Sensor technologies in livestock management	17
Improving piglet welfare	18
Crop biofortification	19
Crop rotations	20
Cover crops	21
Plant growth-promoting rhizobacteria	22
Arbuscular mycorrhizal fungi	23
Nematode pests in agriculture	24
Drought-resistant crops	25
Advances in crop disease detection and decision support systems	26
Mycotoxin detection and control	27
Mite pests in agriculture	28
Supporting cereal production in sub-Saharan Africa	29
Lameness in dairy cattle	30
Infertility/reproductive disorders in dairy cattle	31
Antibiotics in pig production	32
Integrated crop-livestock systems	33
Genetic modification of crops	34

Chapter 1

Defining sustainable forest management (SFM) in the tropics

Francis E. Putz, University of Florida-Gainesville, USA; and Ian D. Thompson, Thompson Forest Ltd.-Kelowna, Canada

1 Introduction

> 'Any claim of sustainable forest management should evoke the queries: What is sus-
> tained? What were the tradeoffs? Over what spatial and temporal scales?'

Sustainable forest management (SFM) is a conceptual codification of forest management practices that continues to evolve from its focus in the 1800s on sustained timber yields. Since the 1987 publication of 'Our Common Future' (also known as the Brundtland Report) by the World Commission on Environment and Development, the definition of SFM has expanded to include the much broader goals of sustaining the economic, social, and environmental benefits from forest. In the words of the United Nations, SFM is a 'dynamic and evolving concept [that] aims to maintain and enhance the economic, social, and environmental values of all types of forests, for the benefit of present and future generations' (FAO, 2018). This broadening of considerations is reflected in the definitions provided by the International Timber Trade Organization (ITTO):

http://dx.doi.org/10.19103/AS.2020.0074.19

[SFM is] the process of managing forest to achieve one or more clearly specified objectives of management with regard to the production of a continuous flow of desired forest products and services without undue reduction of its inherent values and future productivity and without undue undesirable effects on the physical and social environment. (ITTO, 2016)

In this chapter, we elaborate on these definitions of SFM in an effort to promote clarity about the avoidable and unavoidable trade-offs associated with all management decisions; management 'for' something is necessarily management 'against' something else. We also hope to promote measurement, monitoring, and verification of the various indicators of sustainability. It is written out of concern for the obvious deniability of many claims of SFM and the inclusion of vague terms in its definitions such as 'over the long term' without specification of time scales and 'without undue reduction' without clarification of 'undue'. In the chapter we also advocate for clarity about spatial scales and for expansion of the scale at which sustainability is considered from stands up to forested landscapes. Finally, we believe that graphical depictions of the components of SFM, like the one we propose, will help clarify these trade-offs and generally aid our understanding about the challenges of reaching the SFM goal.

Our approach follows the effort of Thompson et al. (2013) to clarify the complex condition of 'forest degradation' through its disaggregation into component biophysical parts. We hope that our un-clustering of the various dimensions of SFM will similarly help inform efforts to promote and evaluate forest management sustainability. We also expand the scope of SFM from individual stands, to which many definitions of SFM pertain, to the scale of forested landscapes, in keeping with other efforts towards the comprehensiveness of land use planning (e.g. Sayer et al., 2016). We hope that our efforts are of use in the development of principles, criteria, and indicators of sustainability for programs such as the Sustainable Landscape Production Certification program under development by the Landscape Standard Consortium (https://verra.org /project/landscape-standard/).

We proceed in this effort to clarify landscape-scale SFM by defining its principal components and then considering them at different spatial and temporal scales. We strive for measurability and precision, in recognition of the diversities of landscapes with forests, characteristics of managed forests and forest managers, forest management goals, and trade-offs associated with land-use interventions. We nevertheless recognize that any definition of SFM with wide applicability and acceptability must be somewhat vague and mutable. That said, whatever the definition of SFM that is adopted, clarity and measurability should be fundamental objectives.

2 Evolving concepts of sustainability

Since well before the Brundtland Report (1987), economists have grappled with what is meant by 'sustainable' and 'sustainability' (e.g. Solow, 1956). While the concerns of foresters about sustainability date back many centuries (reviewed by Wiersum, 1995), the focus was historically on sustaining timber yields, with non-diminishing yields being the goal of management. This focus, which remains relevant, is now referred to as 'strong' sustainability (reviewed by Luckert and Williamson, 2005). In contrast, 'weak' sustainability allows for the transfer of natural capital (e.g. timber stocks or biodiversity) for economic, built, social, and human capital as long as the overall sum of these five forms of capital does not decline. Recognition of the embeddedness of managed forests in landscapes of various other forest and non-forest land uses is more recent, and is reflected in what are known as landscape-level and jurisdictional approaches to sustainability (e.g. Sayer et al., 2016; Stickler et al., 2018; Runting et al., 2019; Griscom et al., 2019.

Expansions of SFM's scope were unavoidably accompanied by modification of the definition of 'sustainability' from one that requires non-diminishing supplies to one that is much more multi-dimensional and negotiable. One consequence of this broadening of the definition and the allowance for capital transfers is that it allows claims of 'sustainable development' and 'sustainable infrastructure'. The expansion of the concept of 'sustainability' to non-renewable resources, as exemplified by the *Journal of Sustainable Mining*, suggests that 'sustainable' is now just a synonym for 'responsible' or 'good' (Putz, 2018).

Here we consider a definition for SFM in the realm of tropical forests that accounts for multiple classes of managed, exploited, and unmanaged forests across landscapes that can include protected areas, selectively logged natural forests, logged forests subjected to additional silvicultural treatments to increase stocking and growth of commercial species, plantations, and forest restoration areas (Fig. 1). We also separate out for consideration forests under the control of rural, local, and/or indigenous communities in full recognition that their lands may host any of these sorts of management practices. Our approach to SFM differs from multiple-use forest management, which typically focuses on compromising goals in areas subjected to similar treatments in what has become recognized as 'land-sharing' (e.g. Phalan et al., 2011). Our approach also expands the 'triad' concept of Messier et al. (2009) in which the focus is on natural forest management, plantation forestry, and forest protection by additionally considering community-based forest management and forest restoration. We hope to shed light on the various benefits derived from different portions of landscapes with forests. We strongly recommend that this approach be expanded by the inclusion

Division of forest landscape for SFM:

1. Protected areas ▪
   - variables: size, location, connectedness, % area
2. Extensive forest management ▫
   - variables: values to achieve sustainably, % area
3. Plantation forest management ▦
   - variables: size of plantations, species, purpose, % area
4. Community forest management ▦
   - variables: values important to community, % area
5. Forest restoration ▦
   - variables: values to achieve sustainability; % area

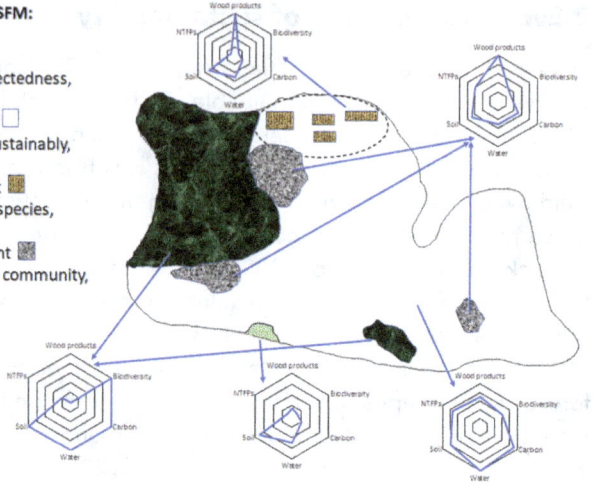

Figure 1 Partitioning a forest landscape for assessment of SFM (extensive and intensive natural forest management not differentiated, but the latter should occur in accessible areas such as near main roads). Each category of forest land-use is evaluated on the basis of the same six criteria illustrated by the biophysical resource hexagons. The overall sum of scores is a relative measure of SFM at the landscape scale for a particular year (The concentric lines inside the perimeter of the main polygons refer to % values of the indicators relative to a primary forest baseline; the blue lines are examples of monitored values from one year).

of more values that are social and economic, and consideration of other land-uses.

The criteria on which we focus are wood products, non-timber forest products, soils, water, carbon, and biodiversity. We assume that the aggregate measure of the extent to which these values are maintained is a measure of the degree of SFM. We recognize that this approach to the assessment of SFM is mostly restricted to biophysical attributes affected by intentional forest management and forest resource exploitation, but we do consider the sustainability of profits from the sale of timber and non-timber forest products (see below). Social, cultural, and other sorts of criteria for more complete assessments of SFM should be readily added to this basic system. Another possible modification that deserves consideration is the use of asymmetrical polygons that illustrate differences in emphases on the various values/criteria.

3 Appropriate scales for assessment of SFM

One difficulty with the concept of SFM is the uncertainty about the scale at which it should be assessed. While forestry practices are implemented at stand scales, given the many values of forests and the inherent trade-offs in any stand-level management regime, SFM might more logically be considered at landscape scales (e.g. Vincent and Binkley, 1993; Boscolo, 2000). In other words, stand-scale management cannot maintain everything, everywhere, all the time, nor should it aim to do so. Only at the landscape level can all forest values be sustained over time and space, if managed properly. With our approach, the sum of scores on all the objectives (i.e. axes in the trade-off polygons), weighted by the area of each land-use, represents a landscape-level measure of sustainability for all criteria (i.e. goods, services, water, carbon, biodiversity, recreation, etc.) at one point in time. We suggest that, with the exception of the most intensive short-rotation tree plantations, multiple-use management with multiple hoped-for benefits is likely to be the goal for most portions of managed forest landscapes. We also recognize that the constraints on achieving the goal of multiple forest management are basically the same, and equally as daunting, as those for SFM in the tropics (Sabogal et al., 2013).

Given the diversity of forest management options, each with its own inherent trade-offs, as well as the diversity of forest conditions, a landscape approach seems appropriate as a first step toward figuring out how the undesired outcomes of management can be minimized overall. Explicit recognition of trade-offs among land uses allows increased rationality in their assessment, as opposed to attempts to maintain all values everywhere all the time. Sizes of managed landscapes sufficient for SFM are likely to be dictated by existing natural constraints, negotiation, geography, and politics, but local values may bound all the others at the upper end of the spatial scale. In these cases, landscape size should reflect some value that would unavoidably be depleted or its maintenance rendered uneconomical as a result of managing at too small a scale. For example, rare species of trees or large mammals may require several thousand square kilometers for persistent populations (e.g. Schulze et al., 2005; Wikramanayake et al., 2011). In these cases, we do not suggest managing for the minimum viable population, but rather some upper value that accounts for temporal stochasticity as well as for controlled and uncontrolled exploitation. In other geographies, for SFM to be practicable, a sufficiently large area may be required to provide an adequate economic benefit from a valued resource (Nasi and Frost, 2009; Sabogal et al., 2013).

4 SFM trade-offs at different scales

Management, by definition, requires that when some species, conditions, processes, or values are managed for, some other species, conditions, processes, or values are managed against. In other words, trade-offs are as inherent to the act of management as they are to resource exploitation. In tropical forests, SFM requires consideration of sufficiently large landscapes for all values to persist, enabling sustainable wood production, sufficient ecosystem services for communities, and no losses of species. For this purpose, areas within the landscape used for different purposes are partitioned into use-categories, often with different values or benefits to society (Fig. 1). For example, intensively managed plantations have limited value for biodiversity but high value for commercial wood production, while protected areas are the opposite. These differences in value-maintenance among land-use categories are represented by the blue value lines inside the trade-off polygons; value lines that approach the outer perimeter of the polygon represent value maintenance relative to the primary forest baseline. Indicators for each of the six criteria are landscape-specific and depend on the specific circumstances and the selected objectives for each forest category. For example, a key indicator for protected areas might be elephant population persistence, while for a plantation, indicators might be a certain amount of wood, fuel, or rubber produced per year. We illustrate five forest categories (Fig. 1), but there are others, such as local community conservation areas and private lands, that may deserve consideration. Regardless of the number of forest categories, each can be evaluated on the basis of the same six criteria in full recognition of the different objectives for which different land-use categories are managed. For more complete assessments of SFM, criteria need to be added that capture the social, economic, and additional environmental values. Extension of this approach to non-forest land uses is possible, but will likely involve specification of new evaluation criteria.

5 Defining terms in SFM

To clarify how SFM might be attained and measured, we commence with definitions of forest (as opposed to plantation), management, and sustained yield. Again, we do not believe that our definitions are sacrosanct, but argue that agreed-upon definitions are needed lest discussions of SFM continue to be plagued by vagueness and ambiguity. A limitation of our approach is that the focus is on forest landscapes and excludes land cleared for agriculture, mining, impoundments or other non-forest land uses, even those with substantial tree cover (e.g., urban forests and some agroforestry systems). We also recognize that our focus is principally biophysical, but hope that our approach is sufficiently adaptable to accommodate social and economic considerations.

Forest versus Plantation: Tree-covered landscapes among which forests with natural regeneration are differentiated from plantations in which all future crop trees are planted (Sasaki and Putz, 2009; Putz and Redford, 2010) often with the intention to clear-cut at frequent intervals. This distinction is made in full recognition of intermediate states, such as selectively logged natural forests that are enriched by planting along cleared lines or in felling gaps. We also recognize that the deleterious environmental impacts of intensive plantation management can be mitigated in many ways such as by maintaining natural forest corridors in riparian areas, increasing structural and floristic diversity within stands, and extending rotations (e.g. Dudley, 2005).

Management: Intentional actions are taken with specified goals, to differentiate management from exploitation and its consequence, degradation. Management occurs at multiple scales and intensities, as fitting for different objectives and in recognition of different trade-offs, and includes protection as well as harvesting. For natural forest management, whether conducted by communities or industrial firms, we differentiate low intensity but extensive approaches based on reduced-impact logging, from higher intensity sorts of silvicultural interventions (e.g. liberation thinning and enrichment planting).

Sustained Yields: The topic of sustained yield forestry has received attention for centuries and may seem more straightforward a consideration than biodiversity, aesthetics, or other values, but we believe it is helpful to disaggregate claims of sustained yield for assessment purposes (Fig. 2). Although most of the data about the effects of sequential harvests are for timber, we believe the same situation applies to non-timber products, especially those for which individuals are harvested in their entirety (e.g. rattan palms).

In the more in-depth assessment of sustained yield proposed here, the degree to which volumetric yields are maintained from one harvest to the next is retained but only as one criterion, described by one axis in a trade-off pentagon (Fig. 2). Given the propensity for harvesters to 'high-grade' (i.e. to select the best individuals first), product quality typically declines with each successive harvest (e.g. increased prevalence of crooked, small, hollow, and heart-rotted trees); we capture this trend in another axis in the sustained yield pentagon. Included in this dimension of sustained yield would be changes in wood densities and working properties such as between old-growth timber and that of regenerating stands of fast-growing trees. The similar tendency to harvest the biggest trees first is represented by an axis that reflects the

Sustained Yield

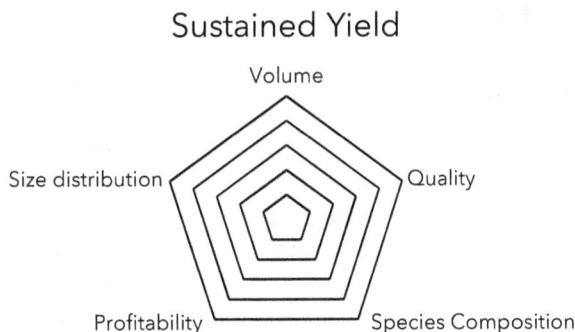

Figure 2 Sustained yield assessed by five criteria. The concentric lines refer to % values of the indicators relative to primary forest (outermost line).

size-class frequency distributions of stems in the post-harvest forest, at a point in time. Given the disproportionately large contributions of very large trees to forest structure, biodiversity maintenance, ecosystem processes, and both population and carbon dynamics (e.g. Lindenmayer et al., 2012; Slik et al., 2013; Sist et al., 2014; Thompson et al., 2014; Kohl et al., 2017), their retention in managed forests is of substantial environmental importance. This importance is reflected in the Brazilian forestry regulation that requires retention of at least 15% of all large trees or the three largest trees per 100 ha harvest block (CONAMA, Resolution no. 1 of 2015; Vidal et al., 2020). Also included in the yield component of SFM is an axis that reflects the sustainability of financial profits. As with all the trade-off polygons, a composite value for sustained yield is calculated as the sum of these five components. Other considerations might, of course, be included and the components might be differentially weighted, but the overall approach provides some clarity about the assessment of claims of sustained yields.

6 Land-use types in SFM

To secure the benefits of landscape-level assessments of SFM, landscapes need to be subdivided into different land-use categories. For a theoretical forested landscape in the tropics, we here consider the following six land-use types:

- 1 Protected Areas;
- 2A Natural Forest Management with Selective Harvests of Timber and Non-Timber Forest Products;
- 2B Natural Forest Management with Silvicultural Treatments After Selective Logging;
- 3 Tree Plantations;
- 4 Community Forests; and
- 5 Forest Restoration Areas.

6.1 Protected areas

Protected areas are designated to maintain ecosystem processes, protect biodiversity and especially low-density species, maintain high carbon stocks, and provide ecosystem services to surrounding landscapes and people. For an example of their importance in a landscape context, one of the few studies that found convergence of conditions in managed forests to those in primary forest noted the importance of associated large protected areas to species and ecosystem process recovery (Norden et al., 2009).

Intact ecosystems in protected primary forests also provide benchmarks against which to measure SFM. Individual protected areas are often not large enough on their own to protect wide-ranging or low-density species. At the landscape scale, proper management of surrounding buffer areas can provide the connectivity among protected areas that is required for persistence of some species (e.g. Hodgson et al., 2011). Like SFM in general, the effectiveness of tropical protected areas is very much a function of governance, stakeholder agreement, level of staff training and commitment, and sufficient funding (Bruner et al., 2001). Assessments of the extent to which protected areas deliver the expected or hoped-for values are made challenging by the tremendous variation in the degree to which protected areas are essentially abandoned or are actively protected with controls on access and resource exploitation.

Unmanaged or primary forests (i.e. those with no visible signs of human intrusion; FAO, 2018) are declining rapidly, especially those that are large. Potapov et al. (2017) reported that globally, the area of intact forest, defined as areas of >500 km^2 with no roads, declined by 7.2% between 2000 and 2013; such areas are already absent in many tropical countries. It is abundantly clear that the stocks of carbon and biodiversity in large primary forests exceed those in forested lands subjected to uses other than protection (e.g. Barlow et al., 2007; Luyssaert et al., 2008; Pan et al., 2011; Edwards et al., 2014; Watson et al., 2018). Many large-bodied and/or heavily exploited tropical animal species prefer intact forests, including the Asian elephant (*Elephas maximus*), African forest elephant (*Loxodonta cyclotis*), tiger (*Panthera tigris tigris*), and harpy eagle (*Harpia harpyja*) (Kinnaird et al., 2003; Barnes et al., 1991; Barlow et al., 2011; Birdlife International, 2016, but see Roopsind et al., 2017). Complicating the discussion of the conservation value of intact forest is the research demonstrating that up to 94% of the area in blocks of forest designated for selective logging remained intact due to the absence of commercial timber, adverse conditions, or poor planning and inadequate supervision (mean = 69%; Putz et al., 2019).

The vast majority of biodiversity exists outside protected areas and the ranges of many species protected partially or predominantly inside parks extend well beyond park boundaries. Hence, protected areas can rarely maintain viable populations of low-density species. Furthermore, in many areas of the world, protected areas either do not exist (Rodrigues et al., 2004) or are unmanaged and subject to illegal activities including poaching and logging (Loveridge et al., 2007; Wittemyer et al., 2008). Laurance et al. (2012) suggested that at least half of Earth's protected areas are failing to sustain their biodiversity. While the rate of loss of intact forests has generally been higher outside areas designated for protection, intact forest areas inside parks nevertheless often decline. For example, Virunga National Park in the Republic of Congo lost 3.3% of its forest cover in just over a decade (Potapov et al., 2017). Overall, the extensively managed forests that serve as buffers for protected areas are essential to sustain tropical biodiversity.

6.2 Natural forest management with selective harvests of timber and non-timber forest products

Natural forests managed for timber and non-timber forest products, which in the tropics typically involves selective harvests, maintain many values when managed properly, as detailed below (Fig. 3). Unfortunately, despite substantial expenditures of time, money, and effort, yields from the harvested species are seldom maintained even when governmental regulations are scrupulously followed (Putz et al., 2012; Vidal et al., 2020). Generally the most valuable timber species are harvested first, followed successively by each of the less valuable ones in subsequent harvests, often referred to as 'logging down the value chain' (e.g. Schaafsma et al., 2013). Furthermore, most government agencies and non-governmental certification bodies (e.g. the Forest Stewardship Council) lack the wherewithal to determine if yields from individual species or even entire forests are maintained (Romero and Putz, 2018).

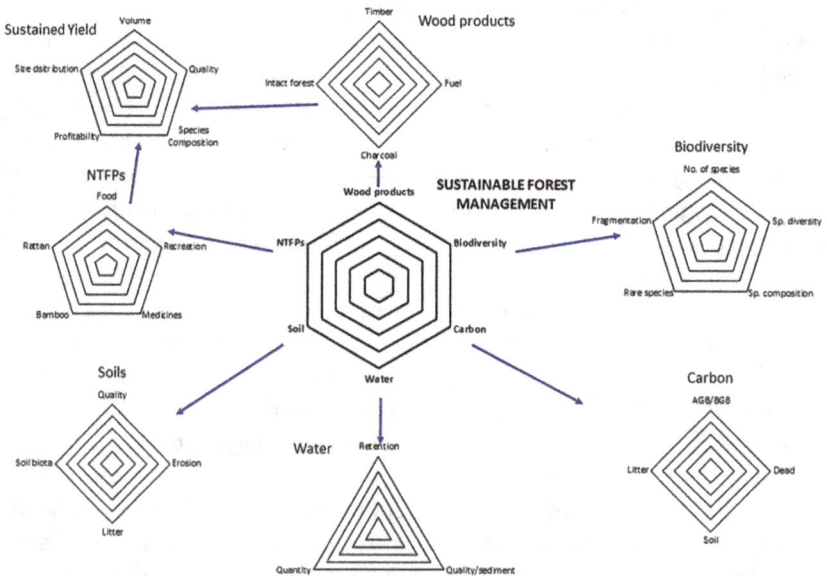

Figure 3 A suggested approach to disaggregation of SFM into its component values with emphasis on biophysical attributes. Note that the criteria and indicators for evaluation of sustained yield pertains to both timber and non-timber forest products. The concentric lines refer to % values of the indicators relative to a primary forest baseline.

Under current regulations in most tropical countries, timber stocks do not regain primary-forest volumes by the end of each officially designated minimum cutting cycle. After conventional timber harvests in Amazonian Brazil, for example, timber volumes take >60 years to recover, not the 25-30 years allowed by law (reviewed by Vidal et al., 2020). Some studies report eventual convergence of logged forests to primary forest conditions (e.g. Norden et al., 2009), while others suggest convergence will not occur (de Avila et al., 2015). A meta-analysis of studies on yield recovery based on >100 publications (Putz et al., 2012) revealed substantial variability, but concluded that timber yields declined by about 46% from the first harvest to the second harvest. That study also reported that, on average, 76% of carbon is retained in forests logged once, and that 85-100% of species of mammals, birds, invertebrates, and plants remain after logging, although long-term persistence is not assured. It is important to note, however, that such studies only report on a few taxa and do not consider all ecosystem functions, especially those delivered by complexes of co-evolved species. Furthermore, the studied forests were not selected at random and were likely representative of the best management underway when the studies were conducted. Finally, although many tropical forests are being logged for the second or third time, most of the reviewed studies focused on timber harvests from primary forest. One consistent message is that, despite the conservation potential of extensive selective logging, SFM is currently jeopardized in much of the tropics by poor logging practices (e.g. Ellis et al., 2019) and premature re-entry logging of previously harvested stands (Sasaki et al., 2016).

The value of extensive production forests for biodiversity and ecosystem services varies with logging intensity (e.g. Burivalova et al., 2014; Franca et al., 2017), logging practices (e.g. Pinard and Putz, 1996; Vidal et al., 2016), but particularly with post-harvest secondary effects including deforestation, poaching, and illegal logging (Michalski and Peres, 2013; Zimmerman and Kormos, 2012; Specht et al., 2015). It is clear that if logging and access are controlled, these secondary effects can be avoided and extensive areas of selectively logged forest will maintain considerable conservation value (Edwards et al., 2011; Putz et al., 2012; Edwards et al., 2014; Lewis et al., 2015; Roopsind et al., 2018, but see Laufer et al., 2013). While extensively managed forests can support much biodiversity, selective logging may nevertheless have substantial deleterious impacts on populations of high-value tree species and associated fauna (e.g. Fisher et al., 2011), partially due to losses of seed sources and dispersal agents. These populations can often be recovered only through carefully managed planting (see 2B).

The ITTO suggested that at least 500 million ha of tropical forest were degraded by 2002; we are not aware of any more recent global estimates of degradation except specifically for carbon (e.g. Baccini et al., 2017). While global attention has swung to reforestation of tree-free areas, restoration of these degraded forests should be a priority. Restoration of degraded forests implies increasing forest resilience, reducing the probability of successful invasions by exotic species, emulating natural processes in silvicultural regimes, and especially avoiding continued degradation. For instance, managing to ensure resilience means maintaining natural species composition and the capacity of forest composition to change under natural circumstances, with only gradual shifts in structure and function (Thompson et al., 2009). Accomplishment of this objective requires that forests are not degraded to a tipping point beyond which the ecosystem state changes radically to a novel and potentially stable condition (e.g. closed forest to open forest). Forest degradation is a difficult concept, however, owing to different perceptions for values derived from the forest, but generally refers to the loss of goods and services (CPF, 2010; Vásquez-Grandón et al., 2018). Degradation becomes easier to measure when considered at the landscape scale by using criteria and indicators (Thompson et al., 2013), like those proposed herein for SFM (Fig. 3) For example, many observers consider plantation to be highly degraded forests or not forests at all (Putz and Redford, 2010), to distinguish them from natural forest. Where plantations replace forests, the environmental losses cannot be recovered, but where plantations are established in already deforested areas, make up only a small proportion of the landscape, are societally accepted, and reduce pressure on natural forests as sources of wood products, they may be acceptable on both environmental and economic grounds. There are also many ways that the deleterious impacts of plantations can be mitigated (e.g. Dudley, 2005), but, based on our observations, few of the recommended practices are ever implemented at industrial scales.

The prerequisite conditions for SFM include proper policy support and legal frameworks, sufficient worker training, uncontested land tenure, sufficient financial incentives, and effective enforcement of regulations (e.g. Nasi et al., 2011; Sabogal et al., 2013; ITTO, 2015, 2016). Lack of financial remuneration for the many environmental services provided by natural tropical forests is one reason for the low financial competitiveness of forest management compared to other land uses such as agriculture and cattle-ranching. To the extent that reduced-impact logging is synonymous with reduced-income logging, it is not reasonable to expect loggers to adopt improved harvesting practices out of enlightened self-interest (Putz et al., 2000). Payments for ecosystem services (PES) seem like a viable mechanism to promote SFM, but successful uses of this tool are scarce and the benefits are ephemeral and funding dependent. It is noteworthy that the 'Socio Bosque' PES program in Amazonian Ecuador reportedly promoted reductions in both deforestation and forest degradation (Mohebalian and Aguilar, 2018). Aside from financial incentives, strong enforcement is also essential to secure the benefits from the vast quantities of carbon that could be sequestered by improved forest management (e.g. Pinard and Putz, 1996; Vidal et al., 2016; Ellis et al., 2019). Halting land grabs and poaching of both wildlife and timber is rendered especially difficult after logging roads improve access and thereby the profits from illegal activities, but the benefits from enforcement are substantial (e.g. Roopsind et al., 2018). Carbon crediting from improved management is sanctioned by the United Nations' Reduced Emissions from Deforestation and Forest Degradation (REDD+) program, but funds remain scarce for promoting the transition from forest exploitation to forest management.

6.3 Natural forest management with silvicultural treatments after selective logging

The principal intervention in tropical forests designated for timber production is selective logging. If properly conducted, selective logging can be considered as a silvicultural technique insofar as it can promote the regeneration and growth of commercial species (e.g. Vidal et al., 2016). Unfortunately, despite decades of promotion of reduced-impact logging (RIL) including millions of dollars spent on RIL policy development and training, most logging still more closely represents timber mining than timber stand management (Ellis et al., 2019). Foresters concerned about logging-induced reductions in timber yields and timber quality, as well as the sequential extirpation of commercial species with each harvest have long prescribed silvicultural interventions. Research firmly establishes the silvicultural benefits of these treatments, but apparently due to insufficient motivation, they are seldom applied outside of research plots.

The toolbox for tropical silviculture includes interventions that range from variations on felling regimes (e.g. strip clear-cuts and group selection harvests), pre-felling treatments such as the cutting of lianas on trees to be felled, as well as the planning of extraction pathways and the marking of trees for directional felling (i.e. RIL). Post-logging silvicultural treatments include liana cutting on future crop trees (FCTs), liberation of FCTs from arboreal competitors, mechanical scarification of felling gaps to promote regeneration, culling of non-commercial trees, and enrichment planting of commercial species along cleared lines or in felling gaps.

High harvest intensities in natural forests typically remove the valuable, mostly shade-tolerant hardwoods, while it damages young recruits, which leads to non-recovery of these species (Van Gardingen et al., 2003; Anitha et al., 2010). Often even low-intensity harvesting can deplete the valuable species (Peña-Claros et al., 2008; Sebben et al., 2008; Schulze et al., 2008a,b, Kukkonen and Hohnwald, 2009), hence the necessity of assisting natural regeneration of over-exploited species, such as with enrichment planting of mahoganies (*Swetenia* spp.), rosewoods (*Dalbergia* spp.), ipê (*Tabebuia* spp.), and cedar (*Cedrela* spp.). The silvicultural effectiveness of each of these treatments is supported by research, but even for these high-value species, few are applied outside of research areas (but see Navarro-Martínez et al., 2017). Historically, more broad-scale employment of silviculture occurred, such as the application of the Malayan Uniform System in Malaysia, but those treated stands were mostly converted to oil palm plantations and silvicultural treatments were discontinued in the forests that remained forest.

For reasons that are not completely clear but that include improved governance and increased recognition of current and pending shortfalls of timber supplies, there are now a few commercial-scale examples of silvicultural intensification of natural forest management (Puettmann et al., 2015). For one, liana cutting on future crop trees (i.e. trees smaller than the minimum cutting diameter that are expected to mature by the end of the cutting cycle) is reportedly more the rule than the exception in a major logging concession in Belize (Mills et al., 2019). Another example is in Indonesia where at least one logging concession carries out large-scale enrichment planting along cleared lines through twice-logged forest (Ruslandi et al., 2017a).

6.4 Tree plantations

While the area of primary forest in the tropics declines, the area of tree plantations (which we do not consider forest *sensu latu* because of the limited species composition, rapid turnover, and usually single objective for wood fiber; Putz and Redford, 2010) increased dramatically over the past two decades. Planted forests now cover >278 million hectares, increasing from 4% to 7% of the reported total tree-covered area between 2010 and 2015 (Payn et al., 2015). Here, we differentiate assisted natural regeneration of native species in extensively managed natural forests, involving directed post-harvest silvicultural treatments, from intensive plantation forestry. Among the plantations are those under short-term (i.e. fastwood) and longer-term cutting cycles, but most often involve a single species used for utility grade timber, chips and fibers, or fuel (Brockerhoff et al., 2008) and support limited biodiversity. Commonly planted are species of the genera *Acacia*, *Eucalyptus*, and *Pinus*. Our justification for this distinction is that forests with assisted natural regeneration also contain many naturally recruited trees and the planted species would not have recovered naturally without the intervention (Thompson et al., 2014; Ruslandi et al., 2017b). We note that although the majority of plantations we have observed in the tropics are appropriately considered 'green deserts', there is plenty of research demonstrating the benefits of biodiversity-enhancing design and management practices such as mixed species plantings and retention of natural forest along riparian corridors (e.g. Dudley, 2005; Paquette and Messier, 2010; Liu et al., 2018). We also note that the assumption that plantations take the pressure of natural forests (e.g. Sedjo and Botkin, 1997) seems supported under some conditions, but remains to be rigorously tested.

6.5 Community forests

We include community forests as a separate land-use category because, although they may be subjected to many different management practices, we assume that SFM is a principal goal. In some cases, however, community forests can also be fully protected, used for ecotourism purposes, or managed by commercial contractors. In any case, many tropical countries are trying to reduce the deleterious impacts of concession forestry and to redress prior local communities' rights violations by assigning management responsibilities to these constituencies. The ethical appropriateness of retuning land to traditional owners notwithstanding, the impacts of community forestry range from relative successes insofar as management improved to failures with rates of deforestation that do not differ from other forests (Bowler et al., 2012; Santika et al., 2017). These failures reportedly resulted from a combination of a lack of training, insufficient funding, disinterest by government in reviewing progress, lack of agreement and coherence of action among community members, land-grabbing, and various illegal/informal activities. In contrast, Porter-Bolland et al. (2012) found that 33 community forests generally had lower rates of deforestation than 40 protected areas, but the mechanisms responsible for this environmental benefit could not be specified due to lack of clear counterfactuals (i.e. what would have happened in the absence of community land tenure). Furthermore, if over time communities accumulate capital and increase in market integration, land-use practices may intensify especially if they are allowed to sell or lease their land.

6.6 Forest restoration areas

Given global attention to the potential benefits of forest restoration (e.g. Griscom et al., 2017), we include this land use, but with some misgivings. One cause of concern is that many of its proponents fail to distinguish between plantations and forests, so that the result of reforestation interventions can differ fundamentally. It is also often unclear whether forest products can be harvested from the reforested areas. Some projects do aim for full ecological restoration, which means recovering the species diversity and composition of primary forest, but it is not clear that this ambitious goal is attainable. Finally, differences in starting conditions affect the outcomes of restoration interventions. For example, the likely outcomes of forest restoration differ between areas that were deforested and then plowed, planted, fertilized, or overgrazed from those that suffered only a clear-cut. Spatial scale and landscape settings also matter, especially if propagules for regeneration need to be dispersed to great distances. In any case, restoration efforts are generally new and of limited overall consequence for the landscape, at least by the year 2020. Worst of all, when naturally tree-poor savannas and grasslands are afforested, the biodiversity consequences are grave (e.g. Veldman et al., 2015).

7 Challenges for SFM in the tropics

To various extents, tropical landscapes present a special case for the implementation of SFM that reinforces the need for disaggregated approaches to assessment, like the one proposed herein. First of all, many forested areas in the tropics are characterized by weak governance, contested land ownership, poverty, large numbers of forest-dependent people, rapid rates of exploitation and forest conversion, modest-to-high opportunity costs of forest retention, and/or political conflicts.

Considerations of sustainability are further complicated by the fact that far more wood is taken for fuel than for timber. For example, Sprecht et al. (2015) found that annual demand for fuelwood by 210 municipalities in Amazonian Brazil was about 300 thousand tons, which they noted would require the clearing of 1200–2100 ha of forest.

Efforts at SFM often face challenges related to the legacies of former interventions: many of the forests exploited for timber today were previously logged, either legally or illegally, but virtually always with little regard for the future. Even in forests with no recent history of exploitation, given high species diversity, tropical trees that produce commercial timber are generally scarce and patchily distributed, which can lead to their rapid commercial extirpation. Many such species, including rosewoods and mahoganies, are now listed by CITES (Convention on International Trade in Endangered Species of Wild Fauna and Flora). Growth and regeneration rates are such that, to maintain viable populations of many commercial timber species, large areas and low-harvest rates are required, at least unless silvicultural interventions, such as liberation thinning around future crop trees and enrichment planting, are applied. Complicating matters further is the fact that some of these species require interior forest conditions and are light, moisture, and thermally sensitive. Furthermore, many tree species depend on co-evolved relationships for pollination, seed dispersal, and nutrient acquisition (e.g. Lewis, 2009; Campos-Arceiz and Blake, 2011).

Given the widespread conversion of lowland forest on gentle terrain to more intensive land uses, natural forest management is increasingly relegated to lands less suitable for industrial agriculture or plantation forestry due to remoteness, nutrient impoverishment, steepness, or poor drainage (e.g. Putz et al., 2018). Remoteness generally increases the likelihood of governance failures while the adverse site conditions render forest lands more susceptible to soil damage and erosion. With the increased intensities of rainfall due to global climate change, soil compaction and erosion, including landslides, also increase, especially on steep slopes (Lele, 2009). The connections between these abuses in the hinterlands and downstream flooding need to be emphasized to spur improved enforcement of land-use regulations even in remote areas.

We recommend that to foster retention of the renewable natural resources and ecosystem services provided by tropical forests, the various values of forests should be disaggregated, considered individually, and then combined in an explicit manner to provide an overall evaluation of the sustainability of forest use at landscape scales. Increased transparency about the trade-offs associated with management decisions at stand up to landscape scales will at least inform debates. To increase the likelihood of political and behavioral changes that lead to improved fates of tropical forests, we advocate for the collaborative construction of detailed and place-specific theories-of-change in which the assumptions are enumerated, relevant actors are identified, their motivations and interactions are captured, and the contexts in which decisions are made are elucidated.

8 Ways forward

One major impediment to sustainable forest management at landscape scales is lack of appropriately trained foresters with the political wherewithal to have their voices heard. This deficiency increases as the number of forestry schools declines almost everywhere partially due to the demonization of tree cutting, despite the canonization of tree planting. Although people will always need wood and wood products, support for improved forest management by international organizations is likewise weak. While the abuses tropical forests suffer from timber mining operations are scrutinized by researchers, few and mostly naïve solutions are offered due to inattention to the relevant factors and constraints. Forest owners, be they governments or communities, also need to forgo some short-term profits so that the renewable natural resources in tropical forests have the chance to be renewed. Perhaps recognition that forest landscapes can be managed sustainably, without denying the many trade-offs, may help efforts to recruit motivated young people into the vibrant field of forestry.

9 References

Anitha, K., Joseph, S., Chandran, R. J., Ramasamy, E. V. and Prasad, S. N. 2010. Tree species diversity and community composition in a human-dominated tropical forest of western Ghats biodiversity hotspot India. *Ecological Complexity* 7(2), 217–24. doi:10.1016/j.ecocom.2010.02.005.

Baccini, A., Walker, W., Carvalho, L., Farina, M., Sulla-Menashe, D. and Houghton, R. A. 2017. Tropical forests are a net carbon source based on aboveground measurements of gain and loss. *Science* 358(6360), 230–4. doi:10.1126/science.aam5962.

Barlow, J., Gardner, T. A., Araujo, I. S., Avila-Pires, T. C., Bonaldo, A. B., Costa, J. E., Esposito, M. C., Ferreira, L. V., Hawes, J., Hernandez, M. I. M., Hoogmoed, M. S., Leite, R. N., Lo-Man-Hung, N. F., Malcolm, J. R., Martins, M. B., Mestre, L. A. M., Miranda-Santos, R., Nunes-Gutjahr, A. L., Overal, W. L., Parry, L., Peters, S. L., Ribeiro-Junior, M. A., da Silva, M. N. F., da Silva Motta, C. and Peres, C. A. 2007. Quantifying the biodiversity value of tropical primary, secondary, and plantation forests. *Proceedings of the National Academy of Sciences of the United States of America* 104(47), 18555–60. doi:10.1073/pnas.0703333104.

Barlow, A. C. D., Smith, J. L. D., Ahmad, I. U., Hossain, A. N. M., Rahman, M. and Howlader, A. 2011. Female tiger Panthera tigris home range size in the Bangladesh Sundarbans: the value of this mangrove ecosystem for the species' conservation. *Oryx* 45(1), 125–8. doi:10.1017/S0030605310001456.

Barnes, R. F. W., Barnes, K. L., Alers, M. P. T. and Blom, A. 1991. Man determines the distribution of elephants in the rain forests of northeastern Gabon. *African Journal of Ecology* 29(1), 54–63. doi:10.1111/j.1365-2028.1991.tb00820.x.

BirdLife International. 2016. Harpia harpyja. The IUCN RED List of Threatened Species 2016 e.T22695998A93537912.

Boscolo, M. 2000. *Strategies for Multiple Use Management of Tropical Forests: An Assessment of Alternative Options.* CID Working Paper Series.

Bowler, D. E., Buyung-Ali, L. M., Healey, J. R., Jones, J. P. G., Knight, T. M. and Pullin, A. S. 2012. Does community forest management provide global environmental benefits and improve local welfare? *Frontiers in Ecology and the Environment* 10(1), 29–36. doi:10.1890/110040.

Brockerhoff, E. G., Jactel, H., Parrotta, J. A., Quine, C. P. and Sayer, J. 2008. Biodiversity and planted forests–oxymoron or opportunity? *Biodiversity and Conservation* 17(5), 925–51. doi:10.1007/s10531-008-9380-x.

Brundtland, G. H. 1987. *Our Common Future.* Report of the World Commission on Environment and Development. Oxford University Press, Oxford.

Bruner, A. G., Gullison, R. E., Rice, R. E. and Da Fonseca, G. A. 2001. Effectiveness of parks in protecting tropical biodiversity. *Science* 291(5501), 125–8. doi:10.1126/science.291.5501.125.

Burivalova, Z., Şekercioğlu, C. H. and Koh, L. P. 2014. Thresholds of logging intensity to maintain tropical forest biodiversity. *Current Biology* 24(16), 1893–8. doi:10.1016/j.cub.2014.06.065.

Campos-Arceiz, A. and Blake, S. 2011. Mega-gardeners of the forest-the role of elephants in seed dispersal. *Acta Oecologica* 37(6), 542–53. doi:10.1016/j.actao.2011.01.014.

CPF Collaborative Partnership on Forests. 2010. *Measuring Forest Degradation.* Available at: http://www.fao.org/3/i1802e/i1802e00.pdf.

de Avila, A. L., Ruschel, A. R., de Carvalho, J. O. P., Mazzei, L., Silva, J. N. M., do Carmo Lopes, M. M., Araujo, M. M., Dormann, C. F. and Bauhus, J. 2015. Medium-term dynamics of tree species composition in response to silvicultural intervention intensities in a tropical rain forest. *Biological Conservation* 191, 577–86. doi:10.1016/j.biocon.2015.08.004.

Dudley, N. 2005. Best practices for industrial plantations. In: Mansouran, S., Vallauri, D. and Dudley, N. (Eds), *Forest Restoration in Landscapes: Beyond Planting Trees.* Springer Science, New York, pp. 379-97.

Edwards, D. P., Larsen, T. H., Docherty, T. D. S., Ansell, F. A., Hsu, W. W., Derhé, M. A., Hamer, K. C. and Wilcove, D. S. 2011. Degraded lands worth protecting: the biological importance of Southeast Asia's repeatedly logged forests. *Proceedings of the Royal Society B* 278(1702), 82-90. doi:10.1098/rspb.2010.1062.

Edwards, D. P., Gilroy, J. J., Woodcock, P., Edwards, F. A., Larsen, T. H., Andrews, D. J. R., Derhé, M. A., Docherty, T. D. S., Hsu, W. W., Mitchell, S. L., Ota, T., Williams, L. J., Laurance, W. F., Hamer, K. C. and Wilcove, D. S. 2014. Land-sharing versus land-sparing logging: reconciling timber extraction with biodiversity conservation. *Global Change Biology* 20(1), 183-91. doi:10.1111/gcb.12353.

Ellis, P. W., Gopalakrishna, T., Goodman, R. C., Roopsind, A., Griscom, B., Umunay, P. M., Zalman, J., Ellis, E., Mo, K., Gregoire, T. G. and Putz, F. E. 2019. Climate-effective reduced-impact logging (RIL-C) can halve selective logging carbon emissions in tropical forests. *Forest Ecology and Management* 438, 255-66.

FAO. 2018. *Terms and Definitions FRA 2020.* Forest Resources Assessment Working Paper 188. Rome, Italy.

Fisher, B., Edwards, D. P., Larsen, T. H., Ansell, F. A., Hsu, W. W., Roberts, C. S. and Wilcove, D. S. 2011. Cost-effective conservation: calculating biodiversity and logging trade-offs in Southeast Asia. *Conservation Letters* 4(6), 443-50. doi:10.1111/j.1755-263X.2011.00198.x.

Franca, F. M., Frazão, F. S., Koraski, V., Louzada, J. and Barlow, J. 2017. Identifying thresholds of logging intensity on dung beetle communities to improve the sustainable management of Amazonian tropical forests. *Biological Conservation* 216, 115-22. doi:10.1016/j.biocon.2017.10.014.

Griscom, B., Adams, J., Ellis, P., Houghton, R. A., Lomax, G., Miteva, D. A., Schlesinger, W. H., Shoch, D., Woodbury, P., Zganjar, C., Blackman, A., Campari, J., Conant, R. T., Delgado, C., Elias, P., Hamsik, M., Kiesecker, J., Landis, E., Polasky, S., Putz, F. E., Sanderman, J., Siikamäki, J., Silvius, M., Wollenberg, L. and Fargione, J. 2017. Natural pathways to climate mitigation. *Proceedings of the National Academy of Sciences (USA)* 114, 11645-50.

Griscom, B. W., Burivalova, Z., Ellis, P. W., Halperin, J., Marthinus, D., Runting, R., Ruslandi, B., Wahyudi, B. and Putz, F. E. 2019. Reduced-impact logging in Borneo to minimize carbon emissions while preserving sensitive habitats and maintaining timber yields. *Forest Ecology and Management* 438, 176-85.

Hodgson, J. A., Moilanen, A., Wintle, B. A., Thomas, C. D. 2011. Habitat area, quality and connectivity: striking the balance for efficient conservation. *Journal of Applied Ecology* 48(1), 148-52. doi:10.1111/j.1365-2664.2010.01919.x.

ITTO. 2015. *Voluntary Guidelines for the Sustainable Management of Natural Tropical Forests.* ITTO Policy Development Series No. 20, Yokohama, Japan.

ITTO. 2016. *Criteria and Indicators for the Sustainable Management of Tropical Forests.* ITTO Policy Development Series No. 21, Yokohama, Japan.

Kinnaird, M. F., Sanderson, E. W., O'Brien, T. G., Wibisono, H. T. and Woolmer, G. 2003. Deforestation trends in a tropical landscape and implications for endangered large mammals. *Conservation Biology* 17(1), 245-57. doi:10.1046/j.1523-1739.2003.02040.x.

Kohl, M., Neupane, P. R.and Lotfiomran, N. 2017. The impact of tree age on biomass growth and carbon accumulation capacity: a retrospective analysis using tree ring data of three tropical tree species grown in natural forests of Suriname. *PLoS ONE* 12(8), e0181187. doi:10.1371/journal.pone.0181187.

Kukkonen, M. and Hohnwald, S. 2009. Comparing floristic composition in treefall gaps of certified conventionally managed and natural forests of northern Honduras. *Annals of Forest Science* 66(8), 809. doi:10.1051/forest/2009070.

Laufer, J., Michalski, F. and Peres, C. A. 2013. Assessing sampling biases in logging impact studies in tropical forests. *Tropical Conservation Science* 6, 16-34.

Laurance, W. F., Useche, D. C., Rendeiro, J., Kalka, M., Bradshaw, C. J., Sloan, S. P., Laurance, S. G., Campbell, M., Abernethy, K., Alvarez, P. and Arroyo-Rodriguez, V. 2012. Averting biodiversity collapse in tropical forest protected areas. *Nature* 489, 290-4.

Lele, S. 2009. Watershed services of tropical forests: from hydrology to economic valuation to integrated analysis. *Current Opinion in Environmental Sustainability* 1(2), 148-55. doi:10.1016/j.cosust.2009.10.007.

Lewis, O. T. 2009. Biodiversity change and ecosystem function in tropical forests. *Basic and Applied Ecology* 10(2), 97-102. doi:10.1016/j.baae.2008.08.010.

Lewis, S. L., Edwards, D. P. and Galbraith, D. 2015. Increasing human dominance of tropical forests. *Science* 349(6250), 827-32. doi:10.1126/science.aaa9932.

Lindenmayer, D. B., Laurance, W. F. and Franklin, J. F. 2012. Global decline in large old trees. *Science* 338(6112), 1305-6. doi:10.1126/science.1231070.

Liu, C. L. C., Kuchma, O. and Krutovsky, K. V. 2018. Mixed-species versus monocultures in plantation forestry: development, benefits, ecosystem services and perspectives for the future. *Global Ecology and Conservation* 15, e00419. doi:10.1016/j.gecco.2018.e00419.

Loveridge, A. J., Searle, A. W., Murindagomo, F. and Macdonald, D. W. 2007. The impact of sport-hunting on the population dynamics of an African lion population in a protected area. *Biological Conservation* 134(4), 548-58. doi:10.1016/j.biocon.2006.09.010.

Luckert, M. and Williamson, T. 2005. Should sustained yield be part of sustainable forest management? *Canadian Journal of Forest Research* 35(2), 356-64. doi:10.1139/x04-172.

Luyssaert, S., Schulze, E. D., Börner, A., Knohl, A., Hessenmöller, D., Law, B. E., Ciais, P. and Grace, J. 2008. Old growth forests as global carbon sinks. *Nature* 455(7210), 213-5. doi:10.1038/nature07276.

Messier, C., Tittler, R., Kneeshaw, D. D., Gélinas, N., Paquette, A., Berninger, K., Rheault, H., Meek, P. and Beaulieu, N. 2009. TRIAD zoning in Quebec: experiences and results after 5 years. *The Forestry Chronicle* 85(6), 885-96. doi:10.5558/tfc85885-6.

Michalski, F. and Peres, C. A. 2013. Biodiversity depends on logging recovery time. *Science* 339(6127), 1521-2. doi:10.1126/science.339.6127.1521-b.

Mills, D. J., Bohlman, S. A., Putz, F. E. and Andreu, M. G. 2019. Liberation of future crop trees from lianas in Belize: completeness, costs, and timber-yield benefits. *Forest Ecology and Management* 439, 97-104. doi:10.1016/j.foreco.2019.02.023.

Mohebalian, P. M. and Aguilar, F. X. 2018. Beneath the canopy: tropical forests enrolled in conservation payments reveal evidence of less degradation. *Ecological Economics* 143, 64-73. doi:10.1016/j.ecolecon.2017.06.038.

Nasi, R. and Frost, P. G. H. 2009. Sustainable forest management in the tropics: is everything in order but the patient still dying? *Ecology and Society* 14(2), 40. Available at: www.ecologyandsociety.org/vol14/iss2/art40.

Nasi, R., Putz, F. E., Pacheco, P., Wunder, S. and Anta, S. 2011. Sustainable forest management and carbon in tropical Latin America: the case for REDD+. *Forests* 2(1), 200–17. doi:10.3390/f2010200.

Navarro-Martínez, A., Palmas-Perez, A. S., Ellis, E. A., Blanco Reyes, P., Vargas Godínez, C., Iuit Jiménez, A. C., Hernández Gómez, I., Ellis, P., Álvarez Ugalde, A., Carrera Quirino, Y. G., Armenta Montero, S. and Putz, F. E. 2017. Remnant trees in enrichment planted gaps Quintana Roo, Mexico: reasons for retention and effects on planted seedling growth. *Forests* 8, 272. doi:10.3390/f8080272.

Norden, N., Chazdon, R. L., Chao, A., Jiang, Y. H. and Vílchez-Alvarado, B. 2009. Resilience of tropical rain forests: tree community reassembly in secondary forests. *Ecology Letters* 12(5), 385–94. doi:10.1111/j.1461-0248.2009.01292.x.

Pan, Y., Birdsey, R. A., Fang, J., Houghton, R., Kauppi, P. E., Kurz, W. A., Phillips, O. L., Shvidenko, A., Lewis, S. L., Canadell, J. G., Ciais, P., Jackson, R. B., Pacala, S. W., McGuire, A. D., Piao, S., Rautiainen, A., Sitch, S. and Hayes, D. 2011. A large and persistent carbon sink in the world's forests. *Science* 333(6045), 988–93. doi:10.1126/science.1201609.

Paquette, A. and Messier, C. 2010. The role of plantations in managing the world's forests in the Anthropocene. *Frontiers in Ecology and the Environment* 8(1), 27–34. doi:10.1890/080116.

Payn, T., Carnus, J. M., Freer-Smith, P., Kimberley, M., Kollert, W., Liu, S., Orazio, C., Rodriguez, L., Silva, L. N. and Wingfield, M. J. 2015. Changes in planted forests and future global implications. *Forest Ecology and Management* 352, 57–67. doi:10.1016/j.foreco.2015.06.021.

Peña-Claros, M., Fredericksen, T. S., Alarcón, A., Blate, G. M., Choque, U., Leaño, C., Licona, J. C., Mostacedo, B., Pariona, W., Villegas, Z. and Putz, F. E. 2008. Beyond reduced-impact logging: silvicultural treatments to increase growth rates of tropical trees. *Forest Ecology and Management* 256(7), 1458–67. doi:10.1016/j.foreco.2007.11.013.

Phalan, B., Onial, M., Balmford, A. and Green, R. E. 2011. Reconciling food production and biodiversity conservation: land sharing and land sparing compared. *Science* 333(6047), 1289–91. doi:10.1126/science.1208742.

Pinard, M. A. and Putz, F. E. 1996. Retaining forest biomass by reducing logging damage. *Biotropica* 28(3), 278–95. doi:10.2307/2389193.

Piponiot, C., Rödig, E., Putz, F. E., Rutishauser, E., Sist, P., Ascarrunz, N., Blanc, L.,Derroire, G., Descroix, L., Laurent, C. G., Marcelino, H. C., Honorio Coronado, E., Huth, A., Kanashiro, M., Licona, J. C. and Mazzei, L. Neves d'Oliveira, M., Peña-Claros, M., Rodney, K., Shenkin, A., Rodrigues de Souza, C., Vidal, E., West, T., Wortel, V. and Hérault, B. 2019. Can timber provision from Amazonian production forests be sustainable? *Environmental Research Letters* 14(6), 064014. Available at: https://iopscience.iop.org/article/10.1088/1748-9326/ab195e.

Porter-Bolland, L., Ellis, E. A., Guariguata, M. R., Ruiz-Mallén, I., Negrete-Yankelevich, S. and Reyes-García, V. 2012. Community managed forests and forest protected areas: an assessment of their conservation effectiveness across the tropics. *Forest Ecology and Management* 268, 6–17. doi:10.1016/j.foreco.2011.05.034.

Potapov, P., Hansen, M. C., Laestadius, L., Turubanova, S., Yaroshenko, A., Thies, C., Smith, W., Zhuravleva, I., Komarova, A., Minnemeyer, S. and Esipova, E. 2017. The last frontiers of wilderness: tracking loss of intact forest landscapes from 2000 to 2013. *Science Advances* 3(1), e1600821. doi:10.1126/sciadv.1600821.

Puettmann, K. J., Wilson, S. M., Baker, S. C., Donoso, P. J., Droessler, L., Armente, G., Harvey, B. D., Knoke, T., Lu, Y., Nocentini, S., Putz, F. E., Yoshida, T. and Bauhus, J. 2015. Silvicultural alternatives to conventional even-aged management–what limits global adoption? *Forest Ecosystems* 2(1), 8. doi:10.1186/s40663-015-0031-x.

Putz, F. E. 2018. Sustainable = good, better, or responsible. *Journal of Tropical Forest Science* 30(1), 1–8. doi:10.26525/jtfs2018.30.1.18.

Putz, F. E. and Redford, K. H. 2010. Tropical forest definitions, degradation, phase shifts, and further transitions. *Biotropica* 42(1), 10–20. doi:10.1111/j.1744-7429.2009.00567.x.

Putz, F. E. and Romero, C. 2014. Futures of tropical forests (*sensu lato*). *Biotropica* 46(4), 495–505. doi:10.1111/btp.12124.

Putz, F. E., Dykstra, D. P. and Heinrich, R. 2000. Why poor logging practices persist in the tropics. *Conservation Biology* 14(4), 951–6. doi:10.1046/j.1523-1739.2000.99137.x.

Putz, F. E., Zuidema, P. A., Synnott, T., Peña-Claros, M., Pinard, M. A., Sheil, D., Vanclay, J. K., Sist, P., Gourlet-Fleury, S., Griscom, B., Palmer, J. and Zagt, R. 2012. Sustaining conservation values in selectively logged tropical forests: the attained and the attainable. *Conservation Letters* 5(4), 296–303. doi:10.1111/j.1755-263X.2012.00242.x.

Putz, F. E., Ruslandi, P., Ellis, P. W. and Griscom, B. 2018. Topographic restrictions on land-use practices: consequences of different pixel sizes and data sources for natural forest management in the tropics. *Forest Ecology and Management* 422, 108–13.

Putz, F. E., Baker, T., Griscom, B. W., Gopalakrishna, T., Roopsind, A., Umunay, P. M., Zalman, J., Ellis, E. A., Ellis, P. W. and Ellis, P. W. 2019. Intact forest in selective logging landscapes in the tropics. *Frontiers in Forests and Global Change* 2, 30. doi:10.3389/ffgc.2019.00030.

Rodrigues, A. S., Akcakaya, H. R., Andelman, S. J., Bakarr, M. I., Boitani, L., Brooks, T. M., Chanson, J. S., Fishpool, L. D., Da Fonseca, G. A., Gaston, K. J. and Hoffmann, M. 2004. Global gap analysis: priority regions for expanding the global protected-area network. *BioScience* 54, 1092–100.

Romero, C. and Putz, F. E. 2018. Theory-of-change development for evaluation of Forest Stewardship Council certification of sustained timber yields from natural forests in Indonesia. *Forests* 9(9), 547. doi:10.3390/f9090547.

Roopsind, A., Caughlin, T. T., Sambhu, H., Fragosa, J. M. V. and Putz, F. E. 2017. Logging and indigenous hunting impacts on persistence of large Neotropical animals. *Biotropica* 49(4), 565–75. doi:10.1111/btp.12446.

Roopsind, A., Caughlin, T. T., van der Hout, P., Arets, E. and Putz, F. E. 2018. Trade-offs between carbon stocks and timber recovery in tropical forests are mediated by logging intensity. *Global Change Biology* 24(7), 2862–74. doi:10.1111/gcb.14155.

Runting, R. K., Ruslandi, R., Griscom, B. W., Struebig, M. J., Satar, M., Meijaard, E., Burivalova, Z., Cheyne, S. M., Deere, N. J., Game, E. T., Putz, F. E., Wells, J. A., Wilting, A., Acrenaz, M., Ellis, P., Khan, F. A. A., Leavitt, S. M., Marshall, A. J., Possingham, H. P., Watson, J. E. M. and Venter, O. 2019. Larger gains from improved management over sparing-sharing for tropical forests. *Nature Sustainability* 2(1), 53–61. doi:10.1038/s41893-018-0203-0.

Ruslandi, W., Cropper, W. P. and Putz, F. E. 2017a. Effects of silvicultural intensification on timber yields, carbon dynamics, and tree species composition in a dipterocarp forest in Kalimantan, Indonesia: an individual-tree-based model simulation. *Forest Ecology and Management* 390, 104–18. doi:10.1016/j.foreco.2017.01.019.

Ruslandi, C., Romero, C. and Putz, F. E. 2017b. Financial viability and carbon payment potential of large-scale silvicultural intensification in logged dipterocarp forest in Indonesia. *Forest Policy and Economics* 85, 95–102. doi:10.1016/j.forpol.2017.09.005.

Sabogal, C., Guariguata, M. R., Broadhead, J., Lescuyer, G., Savilaakso, S., Essoungou, N. and Sist, P. 2013. *Multiple-Use Forest Management in the Humid Tropics: Opportunities and Challenges for Sustainable Forest Management*. FAO Forestry Paper No. 173. Food and Agriculture Organization of the United Nations, Rome, and Center for International Forestry Research, Bogor, Indonesia.

Santika, T., Meijaard, E., Budiharta, S., Law, E. A., Kusworo, A., Hutabarat, J. A., Indrawan, T. P., Struebig, M., Raharjo, S., Huda, I., Andini, S., Ekaputri, A. D., Trison, S., Stigner, M. and Wilson, K. A. 2017. Community forest management in Indonesia: avoided deforestation in the context of anthropogenic and climate complexities. *Global Environmental Change* 46, 60–71. doi:10.1016/j.gloenvcha.2017.08.002.

Sasaki, N. and Putz, F. E. 2009. Critical need for new definitions of "forest" and "forest degradation" in global climate change agreements. *Conservation Letters* 2(5), 226–32. doi:10.1111/j.1755-263X.2009.00067.x.

Sasaki, N., Asner, G. P., Pan, Y., Knorr, W., Durst, P. B., Ma, H. O., Abe, I., Lowe, A. J., Koh, L. P. and Putz, F. E. 2016. Sustainable management of tropical forests can reduce carbon emissions and stabilize timber production. *Frontiers in Environmental Science* 4. doi:10.3389/fenvs.2016.00050.

Sayer, J. A., Margules, C., Boedhihartono, A. K., Sunderland, T., Langston, J. D., Reed, J., Riggs, R., Buck, L. E., Campbell, B. M., Kusters, K., Elliott, C., Minang, P. A., Dale, A., Purnomo, H., Stevenson, J. R., Gunarso, P. and Purnomo, A. 2016. Measuring the effectiveness of landscape approaches to conservation and development. *Sustainability Science* 12(3), 465–76. doi:10.1007/s11625-016-0415-z.

Schaafsma, M., Burgess, N. D., Swetnam, R., Ngaga, Y., Ngowi, S., Turner, K. and Treue, T. 2013. Tanzanian timber markets provide early warnings of logging down the timber chain. In: *15th Annual BIOECON Conference, Conservation and Development: Exploring Conflicts and Challenges*, Cambridge, UK, pp. 18–20.

Schulze, M., Vidal, E., Grogan, J., Zweed, J. and Zarin, D. 2005. Madeiras nobres em perigo. *Revista Ciência Hoje* 36, 66–9.

Schulze, M., Grogan, J., Landis, R. M. and Vidal, E. 2008a. How rare is too rare to harvest? *Forest Ecology and Management* 256(7), 1443–57. doi:10.1016/j.foreco.2008.02.051.

Schulze, M., Grogan, J., Uhl, C., Lentini, M. and Vidal, E. 2008b. Evaluating ipê (*Tabebuia*, Bignoniaceae) logging in Amazonia: sustainable management or catalyst for forest degradation? *Biological Conservation* 141(8), 2071–85. doi:10.1016/j.biocon.2008.06.003.

Sebben, A. M., Degen, B., Azevedo, V. C. R., Silva, M. B., de Lacerda, A. E. B., Ciampi, A. Y., Kanashiro, M., Carneiro, F. S., Thompson, I. and Loveless, M. D. 2008. Modelling the long-term impacts of selective logging on genetic diversity and demographic structure of four tropical tree species in the Amazon forest. *Forest Ecology and Management* 254(2), 335–49. doi:10.1016/j.foreco.2007.08.009.

Sedjo, R. A. and Botkin, D. 1997. Using forest plantations to spare natural forests. Environment: Science and Policy for Sustainable Development 39: 14–30.

Slik, J. W. F., Paoli, G., McGuire, K., Amaral, I., Barroso, J., Bastian, M., Blanc, L., Bongers, F., Boundja, P., Clark, C., Collins, M., Dauby, G., Ding, Y., Doucet, J., Eler, E., Ferreira, L., Forshed, O., Fredriksson, G., Gillet, J., Harris, D., Leal, M., Laumonier, Y., Malhi, Y., Mansor, A., Martin, E., Miyamoto, K., Araujo-Murakami, A., Nagamasu, H., Nilus, R., Nurtjahya, E., Oliveira, Á, Onrizal, O., Parada-Gutierrez, A., Permana, A., Poorter, L., Poulsen, J., Ramirez-Angulo, H., Reitsma, J., Rovero, F., Rozak, A., Sheil,

D., Silva-Espejo, J., Silveira, M., Spironelo, W., ter Steege, H., Stevart, T., Navarro-Aguilar, G. E., Sunderland, T., Suzuki, E., Tang, J., Theilade, I., van der Heijden, G., van Valkenburg, J., Van Do, T., Vilanova, E., Vos, V., Wich, S., Wöll, H., Yoneda, T., Zang, R., Zhang, M. and Zweifel, N. 2013. Large trees drive forest aboveground biomass variation in moist lowland forests across the tropics. *Global Ecology and Biogeography* 22(12), 1261–71. doi:10.1111/geb.12092.

Sist, P., Mazzei, L., Blanc, L. and Rutishauser, E. 2014. Large trees as key elements of carbon storage and dynamics after selective logging in the Eastern Amazon. *Forest Ecology and Management* 318, 103–9. doi:10.1016/j.foreco.2014.01.005.

Solow, R. M. 1956. A contribution to the theory of economic growth. *The Quarterly Journal of Economics* 70(1), 65–94. doi:10.2307/1884513.

Sprecht, M. J., Pinto, S. R. P., Albuquerque, U. P., Tabarelli, M. and Melo, F. P. L. 2015. Burning biodiversity: fuelwood harvesting causes forest degradation in human-dominated tropical landscapes. *Global Ecology and Conservation* 3, 200–9. doi:10.1016/j.gecco.2014.12.002.

Stickler, C., Duchelle, A. E., Nepstad, D. and Ardila, J. P. 2018. Subnational jurisdictional approaches policy innovation and partnerships for change. In: Angelsen, A., Martius, C., De Sy, V., Duchelle, A. E., Larson, A. M. and Pham, T. T. (Eds), *Transforming REDD+: Lessons and New Directions*. CIFOR, Bogor, Indonesia.

Thompson, I. D., Mackey, B., McNulty, S. and Mosseler, A. 2009. *Forest Resilience, Biodiversity, and Climate Change*. A synthesis of the biodiversity/resilience/stability relationship in forest ecosystems. Secretariat of the Convention on Biological Diversity, Montreal. Technical Series no. 43, 67pp.

Thompson, I. D., Okabe, K., Parrotta, J. A., Brockerhoff, E., Jactel, H., Forrester, D. I. and Taki, H. 2014. Biodiversity and ecosystem services: lessons from nature to improve management of planted forests for REDD-plus. *Biodiversity and Conservation* 23(10), 2613–35. doi:10.1007/s10531-014-0736-0.

Thompson, I. D., Guariguata, M. R., Okabe, K., Bahamondez, C., Nasi, R., Heymell, V. and Sabogal, C. 2013. An operational framework for defining and monitoring forest degradation. *Ecology and Society* 18(2), 20. doi:10.5751/ES-05443-180220.

Van Gardingen, P. R., McLeish, M. J., Phililips, P. D., Fadilah, D., Tyrie, G. and Yasman, I. 2003. Financial and ecological analysis of management options for logged-over dipterocarp forests in Indonesian Borneo. *Forest Ecology and Management* 183(1-3), 1–29. doi:10.1016/S0378-1127(03)00097-5.

Vásquez-Grandón, A., Donoso, P. and Gerding, V. 2018. Forest degradation: when is a forest degraded? *Forests* 9(11), 726. doi:10.3390/f9110726.

Veldman, J. W., Overbeck, G. E., Negreiros, D., Mahy, G., Le Stradic, S., Fernandes, G. W., Durigan, G., Buisson, E., Putz, F. E. and Bond, W. J. 2015. Where tree planting and forest expansion are bad for biodiversity and ecosystem services. *BioScience* 65(10), 1011–8. doi:10.1093/biosci/biv118.

Vidal, E., West, T. A. P. and Putz, F. E. 2016. Recovery of biomass and merchantable timber volumes twenty years after conventional and reduced-impact logging in Amazonian Brazil. *Forest Ecology and Management* 376, 1–8. doi:10.1016/j.foreco.2016.06.003.

Vidal, E., West, T. A. P., Lentini, M. W., de Souza, S. E. X. F., Klauberg, C. and Waldhoff, P. 2020. *Sustainable Forest Management in the Brazilian Amazon*.

Vincent, J. R. and Binkley, C. S. 1993. Efficient multiple-use forestry may require land-use specialization. *Land Economics* 69(4), 370. doi:10.2307/3146454.

Watson, J. E. M., Evans, T., Venter, O., Williams, B., Tulloch, A., Stewart, C., Thompson, I., Ray, J. C., Murray, K., Salazar, A., McAlpine, C., Potapov, P., Walston, J., Robinson, J. G., Painter, M., Wilkie, D., Filardi, C., Laurance, W. F., Houghton, R. A., Maxwell, S., Grantham, H., Samper, C., Wang, S., Laestadius, L., Runting, R. K., Silva-Chávez, G. A., Ervin, J. and Lindenmayer, D. 2018. The exceptional value of intact forest ecosystems. *Nature Ecology and Evolution* 2(4), 599-610. doi:10.1038/s41559-018-0490-x.

Wiersum, K. F. 1995. 200 years of sustainability in forestry: lessons from history. *Environmental Management* 19(3), 321-9. doi:10.1007/BF02471975.

Wikramanayake, E., Dinerstein, E., Seidensticker, J., Lumpkin, S., Pandav, B., Shrestha, M., Mishra, H., Ballou, J., Johnsingh, A. J. T., Chestin, I., Sunarto, S., Thinley, P., Thapa, K., Jiang, G., Elagupillay, S., Kafley, H., Pradhan, N. M. B., Jigme, K., Teak, S., Cutter, P., Aziz, M. A. and Than, U. 2011. A landscape-based conservation strategy to double the wild tiger population. *Conservation Letters* 4(3), 219-27. doi:10.1111/j.1755-263X.2010.00162.x.

Wittemyer, G., Elsen, P., Bean, W. T., Burton, A. C. O. and Brashares, J. S. 2008. Accelerated human population growth at protected area edges. *Science* 321(5885), 123-6. doi:10.1126/science.1158900.

Zimmerman, B. L. and Kormos, C. 2012. Prospects for sustainable logging in tropical forests. *BioScience* 62, 479-87.

Chapter 2

The scope and challenge of sustainable forestry

Philip J. Burton, University of Northern British Columbia, Canada

1 Introduction

The wild and managed forests of Earth are remarkable for the many ecosystem services they generate, and on which humanity and other forms of life depend. A forest's productive and regenerative potential also makes it the paragon of a renewable natural resource. This potential, if carefully assessed and harnessed, provides the foundation for the conservation of forest values in perpetuity.

Forestry is the management of tree-dominated ecosystems to promote selected values. It is an ancient practice, an applied science and an inherently human-focussed activity. In other words, forest management is ultimately the management of human activities with respect to forests. Forests and the trees that characterize them developed long before the human species came into existence, they have fluxed and waned with changing climates and have recovered or re-organized themselves after numerous natural disturbances. Forests have persisted without us, but we often manipulate them because we have particular expectations of them. It can be argued that our interaction with forests is essentially self-centred, whether that is to harvest natural resources or to protect vistas and species that we deem valuable. In much of the world, 'forestry' is equated with wood production and harvesting, an unnecessarily narrow definition that is contextualized below and elsewhere in this volume.

http://dx.doi.org/10.19103/AS.2019.0057.01

Similarly, it must be emphasized that logging is not forestry: mere resource extraction is not management if it isn't conducted within a framework of stewardship and consideration for future values.

Sustainability is the ability of a system or an attribute to persist. In the context of natural resources and environmental services, sustainability implies that resources will not be depleted and that the natural environment will not be degraded. The discipline of scientific forestry was one of the first to devise a formal approach to assure sustainable resource production. Yet the concept of sustainability can be complex and nuanced, especially when it comes to the multiple expectations we place on systems as diverse and important as forests. Some of these alternative perceptions are discussed below, while subsequent chapters more fully explore recent developments and options in support of the goal of managing forests sustainably.

2 The natural resilience of forests

All the tree species found today are the products of millions of years of evolution and adaptation. The assemblies of plants, animals, fungi and microbes associated with those trees – forest ecosystems as we know them – have taken shape over several thousand years. Neither the species nor the ecosystems can be considered permanent, but trees as a growth form and forests as ecosystems are incredibly persistent wherever the climate is suitable. The climatic envelope for temperate forests is loosely defined as being between 3°C and 20°C mean annual temperature and between 550 mm and 3400 mm of mean annual precipitation. Boreal or taiga forests are found where mean annual temperatures range from −6°C to 4°C and mean annual precipitation falls between 350 mm and 1500 mm (Whittaker, 1975, p. 167). Within those zones, as in most tropical moist climates, the land 'wants' to grow trees, as the competitive and selective advantages are so great for plants to position their photosynthetic apparatus above that of competing plants, and this typically requires leaves supported from woody branches and sturdy wooden trunks. Shade-loving plants, cavity-nesting birds and wood-decaying mushroom species tag along on the shirt tails of the dominant life form, with different combinations of species finding their optimum expression under different climates, topographic positions and soils to make up thousands of distinctively different forest types around the world.

Yet forests and the evolutionary sculpting of their component species are not just the products of climatic constraints. Depending on the location, forests can be subject to a wide range of physical disturbances, including volcanic eruptions, landslides, floods, windstorms, fires and heavy snow loads. Biotic pressures can sometimes build to outbreak proportions too, killing many trees and changing the composition and structure of forests over wide areas as a result of insect defoliators, bark beetles, fungal pathogens or high levels of vertebrate herbivores - many of which prey upon particular tree species or genera and are thus more selective than many abiotic disturbances. Collectively, the characteristic combination, frequency, event area, and severity of biotic and abiotic disturbances are described as a forest's natural disturbance regime (Runkle, 1985). Every natural forest type is thus the product of evolutionary (long-term) and ecological (recent) selection for traits jointly tolerant of the climate, the terrain and the natural disturbance regime.

Because of the pervasive role of disturbances - at one scale and frequency or another - in all forests of the world, and the long history of natural selection, forest species are generally adapted to persist after disruptions characteristic of the natural disturbance regime under which they developed. Where fire is a characteristic part of the landscape and part of the evolutionary backdrop, we often see thick-barked tree species (such as *Pinus maritima* in southern Europe, *P. ponderosa* in western North America) that are able to survive all but the most intense fires; other species protect their seeds in serotinous (such as *P. contorta*) or semi-serotinous (as in *Picea mariana*) cones, with the seeds released by high temperatures quickly regenerating a recently burnt forest stand (Johnson, 1992). Insect outbreaks and fungal epidemics typically kill a particular species or size of tree, allowing the survivors to take up the newly available resources and fill the recently vacated space (Flower and Gonzalez-Meler, 2015). Some strategies, such as the ability to resprout vegetatively after the aboveground portion of a plant is killed, constitute a generalist adaptive strategy that can allow an individual plant to recover after fire, herbivory, landslide or flood. Seed dispersal traits, often in the form of winged or plumed seeds for wind dispersal, or in the form of attractive or adhesive fruits for animal dispersal, assure species persistence elsewhere even if they cannot persist in place. The shifting combination of species that dominate a forest site after disturbance is expressed in the phenomenon we recognize as ecological succession (Prach and Walker, 2011). Even where species turnover is limited, forest recovery after natural disturbances usually goes through recognizable stages of stand initiation or regeneration, crown closure and self-thinning, maturity and the development of canopy gaps due to the scattered death of mature trees, and (if allowed sufficient time) a self-maintaining old growth stage in which most trees originate in gaps rather after stand-level disturbance (Oliver and Larson, 1996).

But forests are not infinitely tolerant of or resilient to the disruptions that nature or humankind throws at them. Climate shifts associated with continental drift, mountain building and glaciation have resulted in the displacement of boreal and temperate forests for millennia. With many species persisting in refugia or colonizing from other less-affected forest types, forests have remarkably re-occupied suitable habitat in a few thousand years wherever climate was suitable (Pielou, 2008). When fires are too frequent, forest land may revert to steppe or tundra vegetation (Jasinski and Payette, 2005). Where mammalian browsing pressure and human foraging for wood is too intense, forests can be degraded to scrub or even desert, as has happened at low elevations in much of the Middle East and around the Mediterranean basin (Vogt et al., 2007; Sands, 2013). Forests exposed to any novel disturbance – such as exotic invasive species, acid precipitation or ionizing radiation – will inevitably go through a period of reaction and adjustment in which sensitive species are lost and others gradually take their place according to their tolerance to the new stress as well as the other background environmental factors and disturbance regime.

3 The evolution of a concept

Humankind has been using wood for fuel and to construct homes and boats since time immemorial. History documents many examples of forests being depleted and lost as human populations rise, not only in response to the demand for wood, but as more and more forest land is cleared to support agricultural production (Sands, 2013). People also have depended on forests to provide non-timber forest products, ranging from edible fruits and mushrooms to bark used for tanning leather and habitat for wildlife hunted for food. As empires grew and industrial development progressed, forests have been exploited for navies and war machines, to provide timbers for mine supports, and to provide fuel for industrial processes such as ceramic production, glass making and metal smelting. When human populations have temporarily declined as a result of plagues or wars, forests have typically recovered on their own over one or two centuries. Even without logging or cultivation, however, the grazing and foraging of domestic livestock (goats, cattle, swine) and browsing by wild ungulates have often constrained forest regeneration and recovery (Sands, 2013; Innes, 2017).

Fuelwood and timber shortages often prompted local authorities to issue decrees to curtail tree felling and wood gathering; in Europe, such restrictions are documented as far back as the fourteenth century (Innes, 2017). Efforts to assure future supplies of desired wood products are expressed through a history of tree planting, thinning and coppicing that extends into antiquity. Medieval nobles and monarchs from China to England also established forest reserves that were off limits to the population at large, and were protected (often with brutal consequences) and managed for wild game, for the hunting pleasure of the aristocracy. While these early efforts at forest conservation represent the birth of forest management, they also epitomize some of the tensions in sustainable management that persist to this day. Where access to forest resources is limited, those tensions are twofold: first, that protecting forests for future use or enjoyment sacrifices desired uses or profits today (Maser, 1994) and secondly, that there are conflicts between centralized (governmental or industrial) control and the desire of local populations for customary access to harvest desired levels of wood, wild meat and other forest products (Innes, 2017).

In the face of widespread deforestation and timber shortages, the revolutionary concept of sustained yield timber management evolved in central Europe in the eighteenth century. Forest managers realized that it was not enough to simply curtail timber harvesting and forest conversion in an ad hoc manner, but that consideration of the overall forest extent and its rates of tree growth and regeneration would allow the estimation of a level of timber harvesting that could – in theory – continue in perpetuity. This principle of *Nachhaltigkeit* (as the sustainability principle was originally articulated in German) remains at the core of sustainable forest management around the world today (McDermott et al., 2010; Schmithüsen and Rojas Briales, 2012). Simply put, harvesting is limited and silviculture (the regeneration and manipulation of forest stands) is promoted so that the volume of wood harvested in a given management area over a specified period of time does not exceed the amount of wood grown in the same area over the same time period (Fig. 1). Its objectives can be met at two different scales: either by maintaining the size structure within an uneven-aged stand so that an abundance of seedlings, saplings and pole-size trees is there to replace the harvesting of mature trees; or by maintaining the age structure of even-aged stands across a forest estate

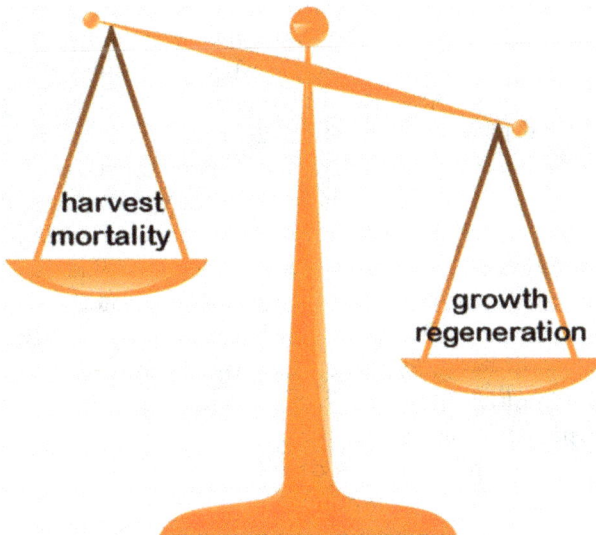

Figure 1 Under sustained yield forestry, timber losses and consumption must be balanced by equivalent (or greater) levels of timber gains through tree growth and regeneration.

as a whole so that younger stands are there to replace the harvesting of mature stands. While these different management approaches have spawned debates over the value of continuous-cover forestry and the use of clearfelling, their appropriate implementation also depends on the size of a forest holding, its management objectives and the silvics (ecophysiological properties) of desired tree species. It can be argued that both approaches merely represent ends of a continuum defined by gap size and forest edge effects, with trees under an uneven-aged management regime harvested in small canopy gaps, and those harvested by clearfelling leaving large gaps in the forest matrix (Coates and Burton, 1997).

Sustained yield forestry was widely adopted on both public and private lands in the northern hemisphere and many European colonies in the nineteenth and twentieth centuries. Often focused on rebuilding timber supplies after wars and other causes of over-cutting, this priority typically resulted in forest transformations from diverse slow-growing native broadleaf species to a few fast-growing (often exotic) conifer species. With its emphasis on fibre production and strict regulation of tree cutting and the need to assure regeneration success, control was not limited to logging, but often included the exclusion of grazing and recreational use, and a general erosion of traditional rights to the commons (Innes, 2017). Expanding acknowledgement of the multiple values of forests and the legitimate rights of various forest users led to the development of 'multiple use' policies in the United States and other temperate jurisdictions in the 1960s. Backlash from rural and Indigenous communities in tropical regions eventually led to the development of a 'community forestry' movement in the 1970s, a concept now adopted in temperate and boreal regions as well, in order to assure better local control over the timber and non-timber benefits of nearby forests (Charnley and Poe, 2007; Gilmour, 2016). The many roles of forests as embraced under the multiple-use and community forest paradigms include the protection of wildlife and fish habitat, the provision of fodder and grazing opportunities, watershed protection and accommodation of recreational activities. In many cases, however, where forest access development and other interventions are undertaken by industrial or government managers intent on commercial wood production, these other roles and values are grudgingly tolerated as constraints, rather than being actively promoted.

The next major steps in the evolution of forestry are associated with a broader societal adoption of the principles of sustainability; indeed, it can be argued that the principle of *Nachhaltigkeit* and sustained yield forestry spawned the concept of sustainability in general (Kuhlman and Farrington, 2010). Following the coining of the term 'sustainable development' by the World Commission on Environment and Development, there is now widespread expectation that interventions in the natural world and investments in human enterprises should '[meet] the needs of the present without compromising the ability of future generations to meet their own needs' (WCED, 1987). Sometimes considered an oxymoron or a politically negotiated compromise (Hauhs and Lange, 2000), the sustainable development mandate nevertheless recognizes the global limits to growth earlier highlighted by the modelling efforts of the Club of Rome (first released in 1972; Meadows and Randers, 2004), and the need for humanity to steward the planet's resources. Sustainable development is often portrayed through the metaphor of a three-legged stool supported equally by pillars of ecological integrity, social acceptability and economic viability (Kidd, 1992). More recent interpretations acknowledge that the economy is a subset of society, both of which are nested within and dependent upon a healthy and productive environment (Fig. 2; Giddings et al., 2002), and that the overriding tension is really between societal consumption and nature's capacity to generate resources (Kuhlman and Farrington, 2010). In the context of forest management, the sustainable development mandate dictates that sustainable forests are not enough, but that sustainable forest communities and sustainable forest enterprises are also necessary ingredients in any recipe for sustainable forestry.

The United Nations Conference on Environment and Development, held in Rio de Janeiro in 1992, resulted in several documents (since endorsed by most nations on Earth) that promoted and advanced concepts of sustainable forestry. Those documents included an overarching set of 27 principles for sustainability (Agenda 21), commitments to save endangered species and their habitats (Convention on Biological Diversity), recognition of the effects of anthropogenic climate change and the potential role of forests in carbon sequestration (Framework Convention on Climate Change) and commitments to combat deforestation and to sustainably manage forest resources under a set of Guiding Principles on Forests (Burton et al., 2003). The promises and aspirations articulated at Rio then spawned several efforts to operationalize sustainability principles into a series of criteria and indicators that could be used to gauge and certify sustainable forest management policies and

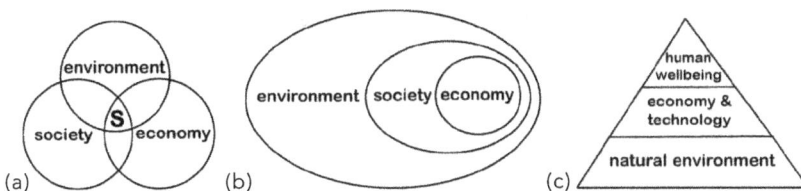

Figure 2 Alternative graphic portrayals of the required components and priorities for sustainable development: (a) the classic Venn diagram that portrays sustainable development, S, at the intersection of ecological integrity, social acceptability and economic viability; (b) an alternative portrayal of economic considerations as a subset of social priorities which in turn depend on a healthy environment; (c) the Costanza et al. (2014) pyramid in which a vigorous economy depends on resources from the natural environment and which is necessary to support social well-being and equity as ultimate goals.

practices. Prompted by consortia of environmental non-governmental organizations, forest products companies and national governments to counteract widespread public protests and boycotts, several certification processes were devised to recognize sustainably produced wood and paper products (Nelson et al., 2003). Those certification criteria, as devised under the Helsinki Process (1993–98) and the Montreal Process (1995–97), endorse the need to maintain productive forests, to protect biodiversity, to protect soil and water, to recognize the contribution of forests to global carbon cycles and to generate socio-economic benefits under appropriate legal and institutional frameworks. These guiding principles now constitute the basis for 'sustainable forest management' (SFM) in many boreal and temperate countries. The SFM paradigm has also been interpreted as extending the sustained yield concept beyond its timber production origins to include all forest values (Adamowicz and Burton, 2003; Higman et al., 2005).

Following the Rio+20 world summit held in 2012, the United Nations adopted 17 Sustainable Development Goals to be implemented from 2015 to 2030. These goals implore all humankind to live within planetary limits, share in a living economy and experience a fair distribution of those benefits (Costanza et al., 2014; see http://www.undp.org/content/undp/en/home/sust ainable-development-goals.html). Sustainable Development Goal (SDG) 15, in particular, calls upon all nations to protect, restore and promote sustainable use of terrestrial ecosystems, sustainably manage forests, combat desertification, halt and reverse land degradation and halt biodiversity loss (Costanza et al., 2014). Sustainable forestry policies and practices are needed to harvest timber without permanently damaging the world's forests, to guide reforestation and afforestation, and to protect forest-dependent wildlife and other forms of biodiversity. It is clear that forests also have a role to play in providing reliable supplies of clean water for human use (SDG 6) through watershed protection. Biomass production in forests also constitutes one of mechanisms for meeting SDG 7, which calls for affordable, reliable and sustainable energy.

Sustainable forestry arguably is positioned to contribute to all 17 of the UN SDGs. For example, by providing meaningful employment in forest planning, harvesting and renewal, and in processing, transporting and marketing wood products, the forest sector provides decent work and contributes to economic growth (SDG 8). With forestry activities often taking place in dispersed and rural locations which often have no other economic activities, the sector also serves an important role in poverty reduction (SDG 1). The wood harvesting and processing industries are constantly innovating, improving efficiencies and releasing new products based on renewable sources of wood fibre (SDG 9). This contributes to their role as responsible producers (SDG 12), with the use of wood in construction contributing to sustainable cities and communities (SDG 11). Forests are increasingly used as the setting for solace and recreation, contributing to good health and well-being (SDG 3). Forest protection, afforestation and prompt regeneration contribute to carbon sequestration and the slowing of climate change (SDG 13). Sustainable and progressive policies in forestry have the potential to lead the way in other SDGs as well. On the other hand, there can be trade-offs and constraints in the pursuit of all SDGs (Nilsson et al., 2016); forests and forest sector investments often must compete with other land uses (e.g. agriculture) and economic sectors (e.g. industrial development) that also contribute to achieving the SDGs.

4 Multiple interpretations of sustainability

Sustainability may be a near-universally accepted goal of modern society, but its interpretation and assumptions vary with every attempt at application. At its simplest, the problem reverts to a question of what resources or values, in what geographic arena, should be sustained? Even though the time horizon is notionally 'in perpetuity', what period of time constitutes a sufficient window for making projections and assumptions – is a decade enough, or even a century? Because we do not want to 'save 'or 'sustain' everything found in the world around us today (including current levels of degraded habitats, pollution, disease and crime), the choice of priorities and their indicators become a value-laden decision. Interpretations of an ideal sustainable world are shaped by cultural history and experience at individual and community levels (Fien and Tilbury, 2002). An underlying tension between the three legs of the three-legged sustainable development stool can be seen in most claims to sustainability, which often favour one of environmental protection, social justice or economic opportunity over the other legs on which we as a society must depend if the Brundtland vision is to be achieved (WCED, 1987).

Some of the key concepts around which sustainability theory, planning and implementation are built include intergenerational equity, perceptions of wealth or values, and the degree to which different values or resources can be considered substitutable (Adamowicz and Burton, 2003). But it may not be enough to assure a diverse, healthy and productive future for one's descendants – intergenerational equity – if those benefits and opportunities are not equally accessible around the world now and in the future. It can be considered irresponsible and elitist if insufficient attention is paid to the 'social' pillar of sustainability, and as long as poverty and deprivation exist within a community, a nation and around the world. For example, it could be argued that many high-income nations maintain high levels of protected and sustainably managed forests on the backs of timber imports from other countries where forest management is unsustainable (Busa, 2013).

In the long run, decisions implemented in the face of social disparity can foreseeably be expected to fail, for poverty, inequality, injustice and exploitation will inspire the underprivileged to disrespect those decisions and plans. At an international level, these situations lead to further disrespect for conservation initiatives and can result in unregulated migration by the desperately poor. People inevitably differ in their conceptions of what is valuable, and whether wealth is interpreted in terms of monetary returns, biological diversity or breadth of recreational and personal development opportunities. This means that the trade-offs necessary in implementing any one sustainability plan will always be unacceptable to one stakeholder or another, because they do not see their values being sustained at the levels desired. Such challenges, though ultimately global in scope, must be addressed by every forest manager with respect to the land and communities for which s/he is responsible. It remains an ongoing challenge for economies, corporations and forest management plans to fully internalize (i.e. account for and fairly pay for) all the external impacts and considerations in their supply chains.

One of the greatest discrepancies in how to implement sustainability can be interpreted as a debate over where policy should aim along the spectrum from 'weak' to 'strong' sustainability. Strong sustainability denotes an emphasis on the conservation of natural capital, with assumptions that many aspects of that wealth (e.g. rare species, primeval forests) cannot recover or be substituted for once they are lost. Weak sustainability, on the other hand, seeks to conserve the overall capital, including built or human and institutional wealth as well as natural resources, and assumes that development, technology and innovation can substitute (or are a fair trade) for the loss of some natural resources (Turner, 1993; Kuhlman and Farrington, 2010). These options are nicely illustrated in the discipline of sustainable forestry, where a strong sustainability emphasis in parks and protected areas demands that managers limit human activities to the fullest extent possible, whereas the development of an industrial or multiple-use forest perceives value in using the revenue from old-growth harvesting to develop a road network and efficient plantations, with the intent of effectively substituting intensively managed second-growth stands for the less productive but greater-volume old-growth forest. So policy makers and the managers they hire have to be quite clear on the management objectives and priorities for any given holding (Noss, 1993): is it sustainable forests (in the strong sense of all-aged ecosystems and all the biodiversity they support), or simply sustainable wood volume production, or sustainable forest management – a spectrum of values including biodiversity, non-timber products, recreation, a viable timber enterprise and community well-being?

5 Challenges in implementing sustainable forestry

Despite being an enshrined principle of responsible forestry for centuries, and even while the concept is being adopted by other sectors of society, it has been challenging to implement and demonstrate sustainable development in the context of operational forest management. Such challenges are understandable for a number of renewable resources – such as wild fish stocks that are difficult to track and annual agricultural crops that are sensitive to vagaries of a single season's weather – but why should it be difficult to manage forests sustainably, when trees and forests are long-lived, stationary and can be readily counted and measured? (Townsend, 2008, p. 190).

Even if trying to simply achieve sustainable fibre (timber) production, one dimension of the problem is that the key elements of wood supply sustainability depend on estimation. Many key parameters and coefficients inherent to wood supply projections and allowable harvest determinations are fluid: estimates of growing stocks, growth rates and regeneration delays; and losses to pests, fires and storms are all subject to error and uncertainty. Year-to-year variation in weather conditions, seed crops and small mammal populations can affect regeneration success (Savage et al., 1996; Kitajima and Fenner, 2000). Tree growth and mortality rates also vary with weather and weather events, and can depend on conditions of stand structure and inter-tree spacing that may be imperfectly characterized and understood. Timber losses to disturbance events – wildfires, windstorms, mass movements, floods, pest outbreaks – can be particularly difficult to anticipate; planning that depends on historical averages can grossly overestimate or underestimate impacts. With climate becoming more variable and disturbances more frequent under the effects of anthropogenic climate change, traditionally used forest yield models are becoming less reliable (Monserud, 2003).

As if those biophysical challenges weren't enough, socio-economic and technological aspects of wood production present even more intractable challenges to sustainability. Oak woodlands historically nurtured to support shipbuilding lost value when ships were instead made of steel (Innes, 2017). The wisdom of planting or promoting one species or group of tree species over another is subject to the whims of market demand, as well as to the vagaries of species-specific outbreaks of native and invasive insect and fungal pests. Many forests reserved or planted for timber are now seen to have more value for amenity purposes such watershed protection, wildlife habitat and recreation, and so are protected from harvesting; this then requires adjustments to the rate of cut that can be sustained elsewhere in the forest estate. Growing human populations and ballooning real estate values mean that considerable forest land and timber production potential is lost to exurban sprawl and residential development, putting further pressure on the timber lands that remain (Drummond and Loveland, 2010). Globalized trading patterns are often able to offset regional differences in fibre supply and demand, but these are sometimes disrupted by tariffs and other trade barriers reflecting the politics of the day. Forest interventions (i.e. forestry and forest restoration) require financial investments, for which there are always alternatives: capital moves worldwide; land may be more valuable for agriculture or residential development; public funds are often diverted to health, education, infrastructure and defence. As forest products enterprises become increasingly concentrated in large, international corporations, investment capital is also increasingly mobile and fickle. Businesses undertaking sustainable wood production that protects the environment, have broad social support, and are economically viable can still be abandoned when profits aren't high enough.

The greatest challenge in implementing sustainable forestry is probably in achieving agreement on precisely what should be sustained, over what area of land, and with what priority to rank the forest values that inevitably conflict (Aplet et al., 1993; Wiersum, 1995; Oliver, 2003). If diverse stand compositions and multiple canopy layers are desired to support avian biodiversity, for example, this may be compatible with non-motorized recreational activities, but at some cost to maximum conifer wood production. Conversely, productive plantations may offer the best opportunity for rapid carbon sequestration, but may have little value for old-growth-dependent vertebrates and lichens. Sometimes a representative balance of these values can be sustained over a sufficiently large landscape through zoning and explicit resource-emphasis management in different zones. This still begs the question of how much land should be allocated to each resource emphasis (each value), and every such allocation decision is subject to our uncertainty as to 'how much is enough' (Tear et al., 2005). In general, the larger a land base available under a given management plan or policy regime, the better the prospects for sustainability of individual values and for the sustainability of net value to society. But large holdings often result in an unbalanced distribution of activities, so that some sites and some local activities bear the brunt of impacts that are clearly not sustainable at a local scale. For example, a village may face damaged vistas, fuelwood shortages and compromised wildlife habitat after rapid industrial logging in easy-to-reach traditional use areas, even though all those values may be sustainable over the geographic and temporal scope of the entire area being managed.

Forestry and its tradition of sustained yield also come with some undesirable baggage that is difficult to jettison or reform. Many forest management policies call for 'maximum sustained yield' and an 'even flow' of timber to sustain mills and forest-dependent employment and rural communities. Maximized yields demand management as close as possible to the feasibility frontier of decision-making, a frontier that is frequently overstepped when disturbances strike or the investment climate changes. The even-flow requirement may inspire a more conservative level of harvesting and assumptions about production, but ignores the event-driven sensitivity of forest ecology, business decision-making and politics. Sophisticated analysis of forest management systems has sometimes pointed out these vulnerabilities, but the response is often to attempt a greater level of command and control, a philosophy of natural resource management with innumerable failures (Holling and Meffe, 1996) and which inevitably inspires protests and backlash from local citizenry. The ultimate irony is that we see examples of forest practices that are in place to make forests follow the assumptions of the models and systems we use to make management decisions, rather than doing the hard work required to make our models and decision-support systems better match the real world.

6 Boosting the sustainability agenda

6.1 Learning by doing

There are many documented examples of temperate and boreal forests being successfully managed for long periods of time, even in the face of the many challenges outlined above. These include planted European forests that have generated multiple generations of timber with limited evidence of any decline in yields over time (Evans, 1999). The plantation wood production model has been successfully exported around the temperate world, and is found, for example, in pine (*Pinus* spp.) plantations in Australia, Chile, New Zealand, South Africa and the US Southeast, as Douglas fir (*Pseudotsuga menziesii*) plantations in Europe and the US Pacific Northwest, and as Chinese fir (*Cunninghamia lanceolata*) plantations in China. Other native forests subject to extensive management (e.g. in eastern Canada) seem to have recovered naturally after being logged or even cultivated (e.g. in the New England states of the United States) a century or more ago, and are once again delivering commercial quantities of timber. Success at sustainable wood production with shorter regeneration delays and shorter rotations has depended on access to sufficient forest land, soil conservation, the planting of nursery-grown seedlings, tree breeding programmes, the control of unwanted vegetation and forest pests, making the more accelerated or intensive forms of forestry dependent on external factors such as publically funded research, the availability and cost of labour, petroleum fuels, chemicals and so on. As noted above, these dependencies on external inputs and subsidizations may not meet everyone's criteria for sustainability.

Whether sustainably produced wood and paper products fully meet our modern definitions of sustainable forest management varies on a case-by-case basis. Even mono-specific planted forests provide habitat for some associated plant, animal and fungal species, and provide watershed protection during the bulk of a rotation (Bauhus et al., 2010). Assessments of the status of various sustainability indicators developed in the wake of the Rio Summit and the Montreal and Helsinki Processes are now being performed on individual management plans (or forest estates) and individual forest policies (at corporate, state or national levels), and are reflected in sustainability certificates and progress reports (Chandran and Innes, 2014). The desirability of certification for sustainably produced forest products has resulted in an expanded consideration of Indigenous and workers' rights, stream protection, biodiversity and carbon sequestration, over and above the direct sustainability of wood or paper products. Hundreds of forest management plans have certifiably implemented the principles and criteria of sustainable forest management on particular forest land bases (Fig. 3). Whether checklists and scorecards according to any one system of criteria and indicators can truly measure progress to sustainability (as claimed by forest products companies and the governments that promote their forest sectors) is an open question, however. Given the youth of these certification programmes relative to the long-term needs of future generations, the sustainability target of these well-intentioned plans remains to be demonstrated. The feedback provided by these evaluations, however, should ostensibly provide incentive for programmes of continual improvement in forest management policies and practices.

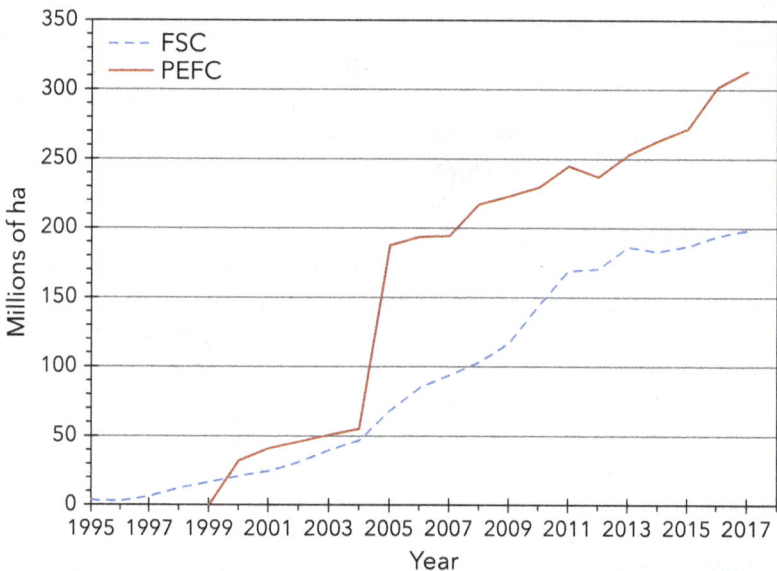

Figure 3 Worldwide growth in forest area (millions of ha) certified as being sustainably managed under the international Forest Stewardship Council (FSC) and Program for Endorsement of Forest Certification (PEFC) systems. Some forest management plans are certified under both systems. Several national-level certification systems (such as the Sustainable Forestry Initiative in the United States, the Canadian Standards Association standards, the Australian Forestry Standard and the Malaysian Timber Certification Scheme) are now recognized by the FSC or PEFC; areas certified by those national systems were gradually incorporated into the numbers shown here over the 2004-08 period. FSC data are from *Facts & Figures* reports and *Annual Reviews* (available from https://ic.fsc.org), and PEFC data are from *PEFC Global Statistics: SFM & CoC Certification Data: December 2017* (available from https://www.pefc.org).

Unfortunately, most of these efforts at implementing sustainable forest management are not well documented in the forestry literature, though many are available on corporate websites and elsewhere on the worldwide web (e.g. Durst et al., 2005). All good sustainable forest management plans (whether certified or not) describe the overall goals for a defined forest area, as informed by the intentions of the landowner and the desires of other interested stakeholders and forest users. Sustainably certified forest management policies, plans and practices can range from a strong emphasis on timber production (e.g. Anon., 2015a) and even plantations of exotic species (e.g. eucalyptus in Uruguay; Anon., 2016), to the protection of old-growth conditions in indigenous forests and the emulation of small-scale natural disturbances (e.g. mixed temperate forest in the US Northeast; Anon., 2014). But most sustainable forest management plans – particularly those on public lands – embrace the legitimacy of multiple values and the rights of multiple interest groups along with some level of timber harvesting, and have struggled with the zoning and prioritization of values within the overall constraints and opportunities presented by the land base available, sensitive environmental features and the economics of management activities (e.g. Anon., 2015b, 2017). Some of the most successful sustainable forest management plans have balanced economically profitable wood production with traditional (Indigenous) land use rights and values (e.g. Davis, 2000; Hammond, 2009). Once again, however, sustainability will be difficult to prove even after a full timber rotation is achieved under such well-intentioned plans – for 'in perpetuity' is a tall order, one that can never be fully verified.

6.2 Forest science in aid of sustainability

Many of the uncertainties and vulnerabilities of sustainable forest management can be reduced through improved scientific knowledge and technological advancements. There is a widespread perception that working with nature, and following natural templates as models for harvesting and silviculture, is a promising avenue for reducing risk and uncertainty (Attiwill, 1994; Gamborg and Larsen, 2003; Long, 2009). So a useful perspective on forest sustainability is attained by reviewing how different disciplines can aid in our understanding and manipulation of forest systems so as to complement the natural resilience of forests and their component processes.

Physiological traits, tolerances and plasticities strongly constrain the survival, growth, health and reproductive potential of tree species and hence the sustainability of wood production and other ecosystem services that depend on a healthy forest cover. These factors are most strongly expressed during stand regeneration and establishment, where abiotic stresses such as drought and cold can be especially pronounced, and where healthy root development of planted stock is particularly important. Innovations in artificial regeneration, nursery production and experience with a wider range of tree species can help assure regeneration success. Net photosynthetic capacity, shade tolerance and light response curves at leaf, branch and whole-plant levels influence the ability to sustain understory plant communities and support efforts to regenerate and promote more multi-storied, multi-species stands. Maintaining healthy soils and nutrient cycles is known to be a key requirement for sustaining forest productivity, but can be a challenge under intensive management regimes. The better we understand the mechanisms, limitations and site- or species-specific aspects of these physiological and ecological processes, the better our chances are for maintaining or improving forest regeneration and productivity.

Genetic diversity and natural selection at multiple levels (individuals, populations, species) have been nature's most important means by which forest species have adapted and evolved in response to massive planetary changes in climate and other factors. Forest scientists recognize the value of protecting and harnessing genetic diversity to fast-track the selection of trees with traits desirable not only for improved wood production, but for resistance to pests and climate extremes too. New techniques of gene sequencing and manipulation can now accelerate the selection and propagation of desired traits even more rapidly, hopefully helping our planted forests stay one step ahead of anticipated stresses.

One of the fundamental challenges of managing for sustainable timber production – indeed for sustainable production of any forest product – is the need for accurate inventories of growing stocks and the accurate monitoring and projection of growth, regeneration and losses due to mortality and disturbances. Aided by new remote sensing technologies (e.g. LiDAR, small unmanned aerial vehicles) and computing techniques, many uncertainties around these fundamental estimates can be reduced. Projecting the future state of the forest – at stand and landscape levels – is more realistic now too, with forest simulation models embracing increasingly more complex stand compositions and structures. Innovations in forest harvesting and engineering are consistently reducing environmental impacts and achieving improved efficiencies, which contribute to the sustainability of both forests and forestry.

Forest biodiversity remains incompletely documented in much of the world, which limits our ability to monitor and conserve it. Other ecosystem services are becoming increasingly important as we strive for sustainable forests and sustainable communities: the role of forests in watershed protection and the generation of reliable, pure stream flows and in taking carbon dioxide out of the atmosphere (where it accentuates the greenhouse effect and contributes to climate warming) and sequestering it in wood and soil come to mind. How all ecosystem services are responding to climate change and how those trends might continue in the future on a site-specific basis is one of the fundamental uncertainties around planning for forest sustainability in the twenty-first century. In general, efforts are being made to shift the forest management paradigm to adopt more diverse and resilient forests that can generate a broader range of goods and services than had been the norm in the past. We need as many new ideas to be proposed and tested as possible if we are to ride out the turbulent changes happening in the biophysical and socio-economic world today.

Natural disturbances and unforeseen causes of widespread tree mortality represent another set of uncertainties nagging away at any attempt to devise a sustainable forest management plan. Efforts to monitor and model insect pests and fungal pathogens depend on improved understanding of their fundamental interactions with host trees and the sensitivity of those interactions to weather and other factors. Innovations in fire science and the ability to predict and reduce the threat of wildfire and the effects of drought and other extreme weather events also have the potential to enhance resilience in temperate and boreal forests, and hence better assure their sustainability.

Much of the financial support needed for the innovative forest management that can effectively address identified conservation and restoration needs depends on the demand for forest products in market-driven economies. New wood products (e.g. engineered wood composites), new processing technologies (e.g. with improved energy and labour efficiencies), new uses for existing products and new markets all contribute to maintaining sustainability of the forest products sector. New technologies and products are emerging all the time, including bio-composites and nano-technology applications. Biofuels from forest biomass have the potential to replace the burning of fossil fuels and further contribute to the mitigation of climate change, while pulp mills can be repurposed to become biorefineries to generate a wide range of products derived from cellulose. Non-timber forest products, including those used for food and for medicine, are not only used on a subsistence level, but have the potential for small-scale sustainable enterprises important to remote communities and often provide leverage for the protection of wild forests. Likewise, planning for forest recreation services and how the desired attributes for that sector can be balanced with wood production and biodiversity protection provides a useful test of the degree to which sustainable forest management has indeed embraced all forest values.

7 Conclusions

Upon reflection, we must concede that sustainable forestry is the aspirational goal of responsible forest stewardship, but not one that is easily attained or demonstrated. With all the estimations, assumptions and uncertainties involved, coupled with a changing climate, shifting social values and unpredictable politics, every sustainable forest management plan is subject to 'shifting goal posts' and becomes a 'living dance' of trying to adjust and adapt to changing realities (Bunnell, 1999). Consequently, we see increasing attention being paid to vulnerability analysis, risk management and resilience planning in all aspects of sustainable forest management (e.g. Eyvindson and Kangas, 2018). By considering the socioecological context of forest ecosystems, the forest products sector and forest-dependent communities as a multi-tiered, multidimensional complex adaptive system, progress can be made in discerning slow processes from fast processes, clarifying the role of positive (accelerating) feedbacks and negative (stabilizing) feedbacks as thresholds are crossed, and in identifying components that are best left to self-organize rather than subjecting them to attempts at control (Filotas et al., 2014). Sustainable forestry is only one piece in the larger jigsaw puzzle of multiple land uses and land use options in most parts of the world, and it is only one cog in the machinery of modern economies with increasingly global connections. The good news is that many regional planning processes, communities, corporations and institutions are wrestling with sustainability goals of their own, for which forestry has a head start and has expertise to share.

Because sustainability implies stasis and the status quo, we must be careful in deciding and articulating what we wish to 'sustain'. We also must acknowledge that the values we choose to protect, conserve and sustain 'in perpetuity' may not be the ones desired by (or feasible for) future generations. In other words, anticipated differences in human populations, socio-economics and environmental conditions may limit our ability to protect, conserve and sustain the values we identify today. A changing climate – with milder winters, more drought, extreme storms, wildfires and pest outbreaks – is further challenging the appropriateness of forest management plans designed to span many decades. Contemporary land use decisions and trees planted today must meet the needs and conditions of today as well as being adapted to climates anticipated for the future, requiring flexible 'no-regrets' decisions. Ultimately, all plans for sustainable forestry are, of necessity, rolling plans that simply maintain some direction of good stewardship and social responsibility in a changing world.

8 Where to look for further information

Additional interesting details on the history and evolution of forestry are provided by Vogt et al. (2007), Sands (2013), Ghazoul (2015) and Innes (2017), sources are described in detail in the References section. Broader discussions on the challenges of sustainable forestry can be found in Aplet et al. (1993), Maser (1994), Oliver (2003), and the entire volumes within which Hauhs and Lange (2003), Adamowicz and Burton (2003), Sands (2013) and Innes (2017) are found. There are currently two main organizations that certify the sustainability of forest products worldwide, namely the Forest Stewardship Council (https://ic.fsc.org/en) and the Program for Endorsement of Forest Certification (https://www.pefc.org/). Both provide standards or criteria and indicators of sustainability with respect to forest management.

Government agencies and academic researchers around the world are actively working to better understand forest ecosystems and to reduce the uncertainties inherent in forest management. In addition to national programmes, two international initiatives that seek to coordinate and communicate forest science research include the European Forest Institute (https://www.efi.int/) and International Union of Forest Research Organizations (https://www.iufro.org/).

9 References

Adamowicz, W. L. and Burton, P. J. (2003), Sustainability and sustainable forest management, in P. J. Burton, C. Messier, D. W. Smith and W. L. Adamowicz (Eds), *Towards Sustainable Management of the Boreal Forest*, pp. 41–64. Ottawa, ON, Canada: NRC Research Press.

Anon. (2014), *Principles of Sustainable Forest Management*, Albany, NH, USA: Tin Mountain Conservation Center. Available on-line at https://www.tinmountain.org/facilities-sanctuaries/timberland-stewardship/principles-of-sustainable-forest-man agement/ (viewed 1 August 2018).

Anon. (2015a), *Sustainable Forestry Policy*, Seattle, WA, USA: Weyerhaeuser Company. Available on-line at https://www.weyerhaeuser.com/application/files/1015/155 4/5217/Sustainable_Forestry_Policy.pdf (viewed 4 August 2018).

Anon. (2015b), *Ecologically Sustainable Forest Management Plan*, Melbourne, Australia: VicForests. Available on-line at http://www.vicforests.com.au/static/uploads/files/ecologically-sustainable-forest-management-plan-v1-0-wfctkbfzjkxi.pdf (viewed 4 August 2018).

Anon. (2016), *Long-term Commitment to Eucalyptus in Uruguay*, Helsinki, Finland: UPM, The Biofore Company. Available on-line at https://www.upm.com/about-us/for-me dia/stories/articles/2016/06/long-term-commitment-to-eucalyptus-in-uruguay/ (viewed 26 December 2018).

Anon. (2017), *Sustainable Forest Management Plan*, Corner Brook, NL, Canada: Corner Brook Pulp and Paper Woodlands, 185p. Available on-line at http://www.cbppl.com /wp-content/uploads/2017/06/SFM_Planv7.pdf (viewed 4 August 2018).

Aplet, G., Johnson, N., Olson, J. T. and Sample, A. (Eds). (1993), *Defining Sustainable Forestry*, Washington, DC, USA: Island Press, 341p.

Attiwill, P. M. (1994), The disturbance of forest ecosystems: The ecological basis for conservative management, *For. Ecol. Manage.*, 63(2–3), 247–300.

Bauhus, J., van der Meer, P. J. and Kanninen, M. (Eds). (2010), *Ecosystem Goods and Services from Plantation Forests*, London, UK: Earthscan, 240p.

Bunnell, F. L. (1999), Setting goals for biodiversity in managed forests, in F. L. Bunnell and J. F. Johnson (Eds), *Policy and Management for Biodiversity in Managed Forests: The Living Dance*, pp. 117-53. Vancouver, BC, Canada: UBC Press.

Burton, P. J., Messier, C., Weetman, G. F., Prepas, E. E., Adamowicz, W. L. and Tittler, R. (2003), The current state of boreal forestry and the drive for change, in P. J. Burton, C. Messier, D. W. Smith and W. L. Adamowicz (Eds), *Towards Sustainable Management of the Boreal Forest*, pp. 1-40. Ottawa, ON, Canada: NRC Research Press.

Busa, J. H. M. (2013), Deforestation beyond borders: Addressing the disparity between production and consumption of global resources, *Cons. Lett.*, 6(3), 192-9.

Chandran, A. and Innes, J. L. (2014), The state of the forest: Reporting and communicating the state of forests by Montreal Process countries, *Int. For. Rev.*, 16(1), 103-11.

Charnley, S. and Poe, M. R. (2007), Community forestry in theory and practice: Where are we now? *Ann. Rev. Anth.*, 36, 301-36.

Coates, K. D. and Burton, P. J. (1997), A gap-based approach for the development of silvicultural systems to address ecosystem management objectives, *For. Ecol. Manage.*, 99, 339-56.

Costanza, R., McGlade, J., Lovins, H. and Kubiszewski, I. (2014), An overarching goal for the UN sustainable development goals, *Solutions*, 5(4), 13-16.

Davis, T. (2000), *Sustaining the Forest, the People, and the Spirit*, Albany, NY, USA: State University of New York Press, 244p.

Drummond, M. A. and Loveland, T. R. (2010), Land-use pressure and a transition to forest-cover loss in the eastern United States, *BioSci.*, 60(4), 286-98.

Durst, P. B., Brown, D., Tacio, H. D. and Ishikawa, M. (Eds). (2005), *In Search of Excellence: Exemplary Forest Management in Asia and the Pacific*, RAP Publication 2005/02, Asia-Pacific Forestry Commission, Bangkok, Thailand: Food and Agriculture Organization of the United Nations. Available on-line at http://www.fao.org/docrep/007/ae542e/ae542e00.htm#Contents (viewed 4 August 2018).

Evans, J. (1999), Sustainability of forest plantations: A review of evidence and future prospects, *Intl. For. Rev.*, 1(3), 153-62.

Eyvindson, K. and Kangas, A. (2018), Guidelines for risk management in forest planning–what is risk and when is risk management useful? *Can. J. For. Res.*, 48(4), 309-16.

Fien, J. and Tilbury, D. (2002), The global challenge of sustainability, in D. Tilbury, R. B. Stevenson, J. Fien and D. Schreuder (Eds), *Education and Sustainability: Responding to the Global Challenge*, pp. 1-12. Gland, Switzerland: Commission on Education and Communication, International Union for the Conservation of Nature and Natural Resources (IUCN).

Filotas, E., Parrott, L., Burton, P. J., Chazdon, R. L., Coates, K. D., Coll, L., Haeussler, S., Martin, K., Nocentini, S., Puettmann, K. J., Putz, F. E., Simard, S. W. and Messier, C. (2014), Viewing forests through the lens of complex systems science, *Ecosphere*, 5(1), art. 1, 23p.

Flower, C. E. and Gonzalez-Meler, M. A. (2015), Responses of temperate forest productivity to insect and pathogen disturbances, *Ann. Rev. Plant Biol.*, 66, 547-69.

Gamborg, C. and Larsen, J. B. (2003), 'Back to nature'–a sustainable future for forestry? *For. Ecol. Manage.*, 179, 559-71.

Ghazoul, J. (2015). *Forests: A Very Short Introduction*. Oxford, UK: Oxford University Press, 150p.

Giddings, B., Hopwood, B. and O'Brien, G. (2002), Environment, economy and society: Fitting them together into sustainable development, *Sust. Devel.*, 10(4), 187–96.

Gilmour, D. (2016), *Forty Years of Community-Based Forestry: A Review of Its Extent and Effectiveness*, FAO Forestry Paper 176, Rome, Italy: Food and Agriculture Organization of the United Nations, 140p.

Hammond, H. (2009), *Maintaining Whole Systems on Earth's Crown: Ecosystem-Based Conservation Planning for the Boreal Forest*, Slocan Park, BC, Canada: Silva Forest Foundation, 389p.

Hauhs, M. and Lange, H. (2000), Sustainability in forestry – theory and a case study, *in* K. von Gadow, T. Pukkala and M. Tomé (Eds), *Sustainable Forest Management*, pp. 69–98. Dordrecht, Netherlands: Kluwer Academic Publishers.

Higman, S., Mayers, J., Bass, S., Judd, N. and Nussbaum, R. (2005), *The Sustainable Forestry Handbook: A Practical Guide for Tropical Forest Managers on Implementing New Standards*, 2nd Edn, London, UK: Earthscan, 332p.

Holling, C. S. and Meffe, G. K. (1996), Command and control and the pathology of natural resource management, *Cons. Biol.*, 10(2), 328–37.

Innes, J. L. (2017), Sustainable forest management: From concept to practice, *in* J. L. Innes and A. V. Tikina (Eds), *Sustainable Forest Management: From Concept to Practice*, pp. 1–32. London, UK: Routledge.

Jasinski, J. P. P. and Payette, S. (2005), The creation of alternative stable states in the southern boreal forest, Québec, Canada, *Ecol. Monogr.*, 75(4), 561–83.

Johnson, E. A. (1992), *Fire and Vegetation Dynamics: Studies from the North American Boreal Forest*, Cambridge, UK: Cambridge University Press, 129p.

Kidd, C. V. (1992), The evolution of sustainability, *J. Agric. Environ. Ethics*, 5(1), 1–26.

Kitajima, K. and Fenner, M. (2000), Ecology of seedling regeneration, *in* M. Fenner (Ed.), *Seeds: The Ecology of Regeneration in Plant Communities*, 2nd Edn, pp. 331–59. Wallingford, UK: CAB International.

Kuhlman, T. and Farrington, J. (2010), What is sustainability? *Sustainability*, 2(11), 3436–48.

Long, J. N. (2009), Emulating natural disturbance regimes as a basis for forest management: A North American view, *For. Ecol. Manage.*, 257, 1868–73.

Maser, C. (1994), *Sustainable Forestry: Philosophy, Science, Economics*, Boca Raton, FL, USA: CRC Press, 400p.

McDermott, C., Cashore, B. W. and Kanowski, P. (2010), *Global Environmental Forest Policies: An International Comparison*, London, UK: Earthscan, 373p.

Meadows, D. and Randers, J. (2004), *The Limits to Growth: The 30-year Update*, London, UK: Routledge, 362p.

Monserud, R. A. (2003), Evaluating forest models in a sustainable forest management context, *For. Biom. Model. Info. Sci.*, 1(1), 35–47.

Nelson, H., Vertinsky, I. B., Luckert, M. K., Ross, M. and Wilson, B. (2003), Designing institutions for sustainable forest management, *in* P. J. Burton, C. Messier, D. W. Smith and W. L. Adamowicz (Eds), *Towards Sustainable Management of the Boreal Forest*, pp. 213–59. Ottawa, ON, Canada: NRC Research Press.

Nilsson, M., Griggs, D. and Visbeck, M. (2016), Policy: Map the interactions between sustainable development goals, *Nature*, 534(7607), 320–2.

Noss, R. F. (1993), Sustainable forestry or sustainable forests? *in* G. H. Aplet, N. Johnson, J. T. Olson and V. A. Sample (Eds), *Defining Sustainable Forestry*, pp. 17–43. Washington, DC, USA: Island Press.

Oliver, C. D. (2003), Sustainable forestry: What is it? How do we achieve it? *J. For.*, 101(5), 8–14.

Oliver, C. D. and Larson, B. C. (1996), *Forest Stand Dynamics*, Update Edn, New York: John Wiley & Sons, 520p.

Pielou, E. C. (2008), *After the Ice Age: The Return of Life to Glaciated North America*, Chicago, IL, USA: University of Chicago Press, 376p.

Prach, K. and Walker, L. R. (2011), Four opportunities for studies of ecological succession, *Trends Ecol. Evol.*, 26(3), 119–23.

Runkle, J. R. (1985), Disturbance regimes in temperate forests, *in* S. T. A. Pickett and P. S. White (Eds), *The Ecology of Natural Disturbance and Patch Dynamics*, pp. 17–33. San Diego, CA, USA: Academic Press.

Sands, R. (2013), A history of human interaction with forests, *in* R. Sands (Ed.), *Forestry in a Global Context*, 2nd Edn, pp. 1–36. Wallingford, UK: CAB International.

Savage, M., Brown, P. M. and Feddema, J. (1996), The role of climate in a pine forest regeneration pulse in the southwestern United States, *Ecoscience*, 3(3), 310–18.

Schmithüsen, F. J. and Rojas Briales, E. (2012). *From Sustainable Wood Production to Multifunctional Forest Management – 300 Years of Applied Sustainability in Forestry*. Working Paper 12/1, International Series Forest Policy and Forest Economics, Zurich, Switzerland: Swiss Federal Institute of Technology, 22p. Available on-line at https://doi.org/10.3929/ethz-a-009955563 (viewed 31 July 2018).

Tear, T. H., Kareiva, P., Angermeier, P. L., Comer, P., Czech, B., Kautz, R., Landon, L., Mehlman, D., Murphy, K., Ruckelshaus, M. H. and Scott, J. M. (2005), How much is enough? The recurrent problem of setting measurable objectives in conservation, *BioSci.*, 55, 835–49.

Townsend, C. R. (2008), *Ecological Applications: Toward a Sustainable World*, Malden, MA, USA: Blackwell Publishing, 346p.

Turner, R. K. (1993), Sustainability: Principles and practice, *in* R. K. Turner (Ed.), *Sustainable Environmental Economics and Management: Principles and Practice*, pp. 3–36. London, UK: Belhaven Press.

Vogt, K. A., Vogt, D. J., Edmonds, R. L., Honea, J. M., Patel-Weynand, T. and Sigurdardottir, R. (2007), Historical perceptions and uses of forests, *in* K. A. Vogt, D. J. Vogt, R. L. Edmonds, J. M. Honea, T. Patel-Weynand and R. Sigurdardottir (Eds), *Forests and Society: Sustainability and Life Cycles of Forests in Human Landscapes*, pp. 1–28. Wallingford, UK: CAB International.

WCED (World Commission on Environment and Development) (1987), *Our Common Future*, Oxford, UK: Oxford University Press, 383p.

Whittaker, R. H. (1975), *Communities and Ecosystems*, 2nd Edn, New York: Macmillan Publishing, 385p.

Wiersum, K. F. (1995), 200 years of sustainability in forestry: Lessons from history, *Environ. Manage.*, 19(3), 321–9.

Chapter 3

The role of certification schemes in sustainable forest management (SFM) of tropical forests

James Sandom, formerly Woodmark Scheme/Responsible Forest Programme – Soil Association, UK

1 Introduction

In the late 1980s, there was considerable concern over forest loss and the deteriorating state of what remained of the world's forest resources, particularly natural tropical forests. These concerns were expressed by scientists, civil society organisations and by local communities, many of whom were directly affected by the changes that the forests were undergoing. The growing consensus was that the existing systems of regulation and control that were intended to maintain forests and ensure their continuity at a national level were simply not working, and the situation was exacerbated by the absence of any meaningful or effective international system of control (Petera and Vlosky, 2006; Cashore et al., 2006; Auld, 2014).

At this time certification was not a concept, or practice, that was widespread in the forestry and timber sectors. Where it did occur, it was usually in the form of a quality management system (QMS) such as the ISO 9000 (Bass, 1996). The ISO 9000 quality management system had been launched in 1987 by the International Standards Organisation (ISO) which used certification as a management tool to help commercial companies achieve a consistent delivery of products and services. The ISO 9000 used a succession of structured audits by which a company could measure its performance against a standardised set of requirements.

Forest certification or, more accurately, the certification of wood and timber products, was proposed as a possible solution that was objectively verifiable and which could be applied globally. Forest certification represented a radically different approach to forest regulation. It avoided the existing 'command and control' models of regulation and substituted a system based on meeting pre-determined performance standards which could be monitored over time through a succession of systematic, objective audits. Key members of the international community thought that a suitably modified certification system could offer the basis for an alternative and viable system of forest regulation by which the sustainable utilisation and conservation of the world's forests could be assured.

The idea of certification as a possible way to achieve sustainable forest management (SFM) was not initially accepted by everyone. Some felt that the two aims of certification - utilisation and conservation - were fundamentally irreconcilable and that any level of commercial exploitation of forest resources could only be damaging for long-term sustainability. It is also important to understand that this potential solution was not initially promoted by the traditional players in the forest and timber industry - such as national forest authorities, regulators or traders - but by international non-government organisations (NGOs).

NGOs were organisations or alliances comprising stakeholders with shared environmental and social aspirations, and by the mid-1980s, they had become a powerful and effective force in the debate over the management of the world's resources. However, many within the forest sector and timber trade regarded NGOs as self-appointed, unrepresentative and anti-establishment in their views. The initial response to any form of certification from the forestry and timber trade's traditional stakeholders was almost uniformly negative.

The NGOs continued to pursue forest certification as a workable solution and from the early 1990s they worked to develop a form of certification that they believed would be effective and which could be accepted by the international community, national governments and commercial producers and processors (Cashore et al., 2004, 2006; Auld, 2014). The NGO's development work resulted in the development of the Forest Stewardship Council (FSC) and made use of the pioneering work of the Rainforest Alliance, which had developed its own Smartwood certification system while working with Indonesian state-owned forests in the late 1980s: this had a significant influence on the design and structure of the certification system developed by the FSC (Newsom and Hewitt, 2005).

The FSC was formally launched in 1993 and its certification scheme became fully operational in 1996. In spite of continuing resistance to certification, which proved particularly strong in countries that traditionally exported tropical timber, forest certification schemes such as the FSC scheme gradually began to be taken seriously and, by the 2000s, they had become accepted as an integral part of commercial forestry and the timber trade.

A number of regional certification schemes and standards were formulated, but few of these developed into certification schemes that were commercially viable or widely adopted. In contrast, standards and certification systems devised for particular tree crops - frequently grown as plantation monocultures in the tropics - were developed and have been adopted much more readily. These include rubber, oil palm and fast-growing species grown for biomass. In the case of oil palm there are now a number of nationally based certification schemes as well as international schemes such as the Roundtable for Sustainable Palm Oil (RSPO). These have been taken up widely by governments and commercial growers to demonstrate the crop's sustainability, but they are controversial, given that many of these crops had been established through widespread land clearance and conversion of the original vegetation cover, which was frequently natural forest. As a consequence, the credibility of these certification schemes has been frequently challenged, especially by environmental NGOs and civil society organisations (CSO) (Gatti et al., 2019).

A number of national schemes and standards were also developed which experienced varying degrees of success and international acceptance until the PEFC scheme evolved and subsumed them within its mutual recognition format: this is now the only forest certification scheme other than the FSC that can be considered truly international in scope. Originally launched in 1999 the Pan-European Forest Certification Scheme had been developed as an alternative to the FSC Scheme and was specifically designed to meet the needs of forest owners in Scandinavia and then Europe.

The scheme was substantially restructured in 2003, when its aspirations became global and it changed its name to better reflect its aspirations and methodology, although it was able to retain its original acronym. The re-badged 'Programme for the Endorsement of Forest Certification' (PEFC) accurately describes the approach adopted by this scheme, where national certification standards and schemes are developed by national stakeholders under the umbrella of the PEFC framework: as a consequence the PEFC has subsequently become the forest certification scheme with the largest area of certified forest.

By the 2000s there were many different schemes available to commercial companies. These had different standards and definitions of sustainability that sought to satisfy different stakeholders and their expectations. Some had been developed as national schemes, which aimed to provide exporters or international customers with a degree of assurance about the source, legality or sustainability of the timber or wood-based products being supplied to the market. Other schemes focussed on a more equitable distribution of revenues to the primary producers. As a result, the certification landscape became progressively more complex, including initiatives and organisations committed to fair trade and business to business (B2B) trading models.

NGOs and commercial consultancies began to offer technical and advisory support services to commercial companies eager to gain access to markets for certified timber. As an example, the World Wide Fund for Nature (WWF) developed its own Global Forest Trade Network (GFTN) which, by offering its own seal of credibility together with preferential (if limited) access to the markets demanding FSC-certified timber, potentially competed with the FSC scheme that they had been instrumental in establishing (http://gftn.panda. org/). An additional complication has been that, with the exception of the PEFC, there has been a reluctance to consider mutual recognition between different schemes (Humphreys, 2006).

The practical result is that forest certification comprises a wide range of different certification systems and structures whose impacts on sustainability and SFM vary widely. An objective assessment of certification, especially in relation to its contribution to delivering SFM, is a difficult and complex exercise, rendered particularly difficult by the paucity of relevant and objective data. To understand how effective certification – as a movement – has been at promoting and delivering SFM, it is vital to understand how certification developed and why the certification industry is structured as it exists today. Only when the underlying influences and motivations are fully understood, will it be possible to attempt an objective assessment of the impact of certification in contributing to SFM, an objective which, even now, remains itself subject to re-interpretation and re-definition.

Forest certification and its impact have proved to be fertile ground for research and study, both political and technical, and a huge number of research articles, papers and books have been accumulated on the subject. This chapter seeks to offer an overview of this research, some of the main conclusions that have already been reached about the impact of forest certification and how much it has been able to contribute to the goal of SFM. Finally, the chapter offers some suggestions as to what modifications could be made to the current structure of global and regional forest certification to ensure that it better achieves this goal.

2 The need for change

By the late 1970s, there was increasing concern among certain members of the international community that the world's forests were in trouble. Objectively verifiable and reliable time series data was limited or restricted to specific geographic areas. Even the FAO, the body responsible for generating global data about forests, was reliant on data supplied by member countries that was often out of date or of uncertain reliability. Nevertheless, a body of data and anecdotal evidence had started to accumulate and indicated that the principal forested areas of the world were in decline due to an unprecedented combination of environmental, political and economic pressures, resulting in increasing levels of deforestation and forest degradation.

In some cases, the root causes were the result of political and economic choices: to convert a forest into an alternative land use, usually commercial agriculture; to exploit a mineral or water resource, or to undertake large-scale infrastructural development. All of these required the forest to be removed. Additionally, there were a number of cases, mainly in the tropics, where wars and conflicts had directly led to the long-term loss of forest cover or its impoverishment. These included extensive forest areas in Vietnam, Laos, Cambodia, Mozambique and Angola. Some types of forest cover were proving particularly vulnerable and suffered disproportionate losses; these included the mangrove forests, many of which were being converted to commercial seafood production. But other types of loss or degradation were less easy to explain, such as the decline of the conifer forests of central and eastern Europe, which had initially been attributed to 'acid rain' but which subsequently proved to be the result of a combination of circumstances.

In order to establish the genuine root causes driving the loss of the forests – and develop appropriate policies and interventions – more reliable and objective data was required together with more in-depth analysis. Fortunately, satellite imagery had started to become available from 1975 (initially Landsat 2 imagery and then subsequent versions). Although interpretation and statistical analysis was still being developed, satellite imagery was used increasingly by national and international agencies to determine the true status of forest resources and establish clear data and baselines. This new source of data and information confirmed that the fears of widespread destruction and degradation of forests in the tropics were essentially correct.

The improved data, together with more robust analysis, demonstrated that there were many drivers that were contributing to the decline in forests. Within the tropics, it was clear that nations were increasingly utilising their natural resources, including their forests: not only were forests being subjected to more intensive harvesting for their timber (commonly regarded as 'natural capital'), but they were frequently being cleared completely, principally to permit the land to be utilised for agriculture, mining or other large-scale infrastructure projects. These processes coincided with a new generation of more efficient machines capable of rapid, large-scale harvesting and clearing, and which were able to penetrate areas that hitherto had been considered inaccessible. Even where large-scale conversion of forest resources was not the declared objective, the fragmentation of forests, the increased access offered by logging roads, and the influx of workers to unpopulated areas all contributed to the progressive elimination or impoverishment of natural forests.

In the 1970s one of the most visible results of this increasing exploitation of forest resources was the increased volume of timber that was becoming commercially available. This resulted in the international timber trade expanding vigorously. Between 1960 and 1980 timber production rose by 25% (from 2.4 billion m³ to almost 3.0 billion m³) (http://www.fao.org/forestry/statistics/). In addition, fundamental shifts occurred in the sourcing of timber and the resulting supply chains. Countries which had previously been considered minor players became major suppliers, generating unprecedented levels of revenue and profit.

Although Malaysia had been a traditional supplier of tropical timber for many years, its trade expanded rapidly in the 1960s and 1970s with increased production from the exploitation of the forests of Sabah and Sarawak. Indonesia, Papua New Guinea, Brazil and the Philippines all became major sources of tropical timber. This timber was not only utilised by the traditional markets of Europe and North America, but increasingly by Japan, Korea, Thailand and other rapidly expanding economies of Asia. The international timber trade became increasingly complex and a number of the supplying countries started to become increasingly dependent on the revenue generated by timber and non-timber forest products.

It was also becoming clear that these economic, social and environmental drivers were overwhelming the command and control measures that had traditionally been employed by national governments to ensure the survival of natural forests or maintain a sustainable utilisation of their products, typically by limiting the levels of harvesting (e.g. by size or species of tree) or restricting harvesting to specific areas. The combination of the high revenues that could be generated, together with the weak or ineffective existing control measures in some countries and a trade that had consistently resisted transparency, provided numerous opportunities for unscrupulous operators to adopt practices that were unsustainable or illegal.

A number of national bodies, as well as some international observers, began to discern a disturbing trend: the establishment of a new form of 'organised crime' based on the exploitation of forests and the illegal trade in timber and forest products. This illegal trade was highly lucrative, international in scope and reinforced by opaque systems of ownership and linkages between those exploiting forest resources, government agencies and personnel. These new ways of operating proved difficult to penetrate and very resistant to policing and reform. The involvement of government officials in this illegal trade in some cases meant they represented a threat to the very fabric of good governance (Kleinschmidt et al., 2016).

If the 1970s were the decade in which the problems inflicted on the world's forests had first become apparent, the 1980s were the period that established the true nature, complexity and scale of the problem. This was facilitated by a combination of factors that included, first and foremost, the expanding volume of hard and incontrovertible data that was being accumulated from an expanding range of sources. Satellite imagery was increasingly available and, rather than being limited to academic institutions, this data was being interpreted and subjected to more rigorous and demanding analysis by a wide range of environmental and social organisations. Time series data was also now available that could track patterns of change and predict future trends with greater precision.

A second important factor was the expansion of organisations – governmental and non-governmental – that were actively involved in, or were seeking to change, the way that the world's forests were being managed. For some organisations, forestry was simply one aspect of a broader goal to which they were committed. The WWF was one such organisation, established in 1961 to promote conservation of nature, including conserving and protecting the world's natural forest ecosystems. By the 1980s, it had become one of the largest, independently funded and most powerful NGOs and its forestry programme was well-established and active globally in raising awareness of the plight of the world's forests development and the development of practical solutions. These NGOs brought an enhanced effectiveness to advocacy and they also undertook practical action, using funds provided by an informed and committed membership. These actions included monitoring, oversight, political lobbying and new implementation strategies with practical action including, in some cases, the purchase of forest resources.

Examples of other prominent NGOs that became involved in tropical forestry included: Friends of the Earth (1969); Greenpeace (1971); World Resources Institute (1982); Environmental Investigation Agency (1984); Conservation International (1987); World Rainforest Movement (1982); the World Land Trust (1989); and Forest People's Programme (1990). These are just a small sample of the international and national NGOs that arose during this period, joining an older generation of environmental NGOs such as the Sierra Club (1892); The Nature Conservancy (1951) and Flora y Fauna (1903). All these organisations continue to be actively involved in trying to resolve the problems facing the world's forest resources. Although frequently targeting different approaches and solutions, the effectiveness of these new international NGOs helped transform the global debate on sustainability in general and forest sustainability in particular.

During this period a number of multilateral government organisations were also trying to address the problems that had already been identified. The UN FAO had long been an active player in the international forestry sector as a source of technical expertise and project funding. The World Bank had developed a significant and influential portfolio of international forestry projects around the world that often supported, or worked closely with, the bilateral aid programmes of many western-developed countries, including the United States, Canada, Sweden, France, Germany, Japan and the United Kingdom.

In 1985 the World Bank and the FAO were instrumental in launching the Tropical Forest Action Plans (TFAPs), an ambitious programme for channelling funding from developed to developing countries by identifying appropriate investment opportunities, particularly for countries possessing tropical forests (Burley, 1988). The TFAP agenda became progressively contentious, particularly among environmental and social NGOs, which became increasingly concerned about the potentially negative impact of some of the investment strategies that were developed in relation to commercial development of forest resources. The critical reaction to the TFAP contributed to a major change of World Bank policy in the early 1990s and greater emphasis on incorporating sustainability issues (environmental and social) into its development programmes (Winterbottom, 1995).

NGOs also questioned the effectiveness of the International Tropical Timber Organisation (ITTO), an intergovernmental organisation established in 1986 as the body responsible for overseeing and managing the markets for tropical timber. The membership comprised representatives of most of the countries responsible for supplying tropical timber, together with representatives of the major users. It was effectively a trade organisation and the initial aims of the organisation were vague on the key issues of conservation and sustainability.

Many international NGOs questioned the fundamental rationale of such an organisation and, although its mandate was refined through subsequent revisions of its founding document (the International Tropical Timber Agreement), where the goals of sustainability and conservation were increasingly referenced, the ITTO was accused of failing to establish any practical or effective measures to help bring them about (Auld, 2014). Although the ITTO was considered one of the principal institutions through which forest certification could have been implemented, the membership was initially unwilling to consider certification and failed to reach a consensus on developing a timber labelling system that they feared might restrict the trade in tropical timber. This lack of concrete action reinforced the initial scepticism of the NGO community and by 1993 many perceived the ITTO as part of the problem, rather than an organisation that could offer viable solutions (Humphreys, 1996, 2006).

Another option that was available but was not actively pursued by the wider NGO community, was the establishment of a multilateral convention for forestry and the international trade in timber. A Convention on the International Trade in Endangered Species (CITES) had been established in 1975 to regulate the trade in endangered species, and it had operated with some success for over 10 years. A similar global convention for forests and the timber trade was suggested as a possible solution, obliging signatory nations to develop and adopt sound and sustainable management practices for their forests. The advantages of a convention were that it would be global in scope and would be binding on all signatory parties. However, many NGOs were reluctant to initiate a process which would take years to reach fruition (the CITES process had taken 12 years).

A convention would require considerable negotiation and the final terms would inevitably be a compromise. NGOs feared that, even if a suitable convention could be agreed, the result might not be sufficiently rigorous to achieve their desired goals. The lesson from the International Convention on the Regulation of Whaling (1946) was that quotas were very difficult to police effectively and that the only really effective regulatory measure was a global ban.

However, unlike whales, which were regarded as a global asset that belonged to all rather than one specific nation, forests were considered very much a legitimate national asset to be utilised to generate revenue and promote national development. These concerns meant NGOs were slow to exploit the potential of the international convention adopted for the conservation of biodiversity: the Convention on Biological Diversity launched in 1993. It took only 5 years to finalise and has subsequently become a cornerstone of international cooperation. It includes a wide range of internationally agreed protocols, particularly on biosafety but does not directly address forest management.

Another reason NGOs were sceptical about the effectiveness of a global convention was the countervailing influence of the World Trade Organisation (WTO), the global international governance body responsible for regulating trade (Klabbers, 1999). The WTO had started the Uruguay round of trade negotiations in 1986. This engaged the 123 member nations for 8 years of negotiation as they sought to extend the scope of WTO to cover a range of new products and services, including agriculture. The WTO contained mechanisms designed to enable it to challenge any activity considered a barrier to trade, either in the form of a tariff or, more controversially, any other action that could be considered a non-tariff trade barrier. In the absence of any in-built environmental assessment criteria, this gave rise to concerns that any attempts by a country to restrict trade on environmental grounds might be seen by others as a non-tariff barrier subject to significant WTO financial penalties.

Two cases, relating to restricting fishing of tuna on environmental grounds, were found by the WTO to be discriminatory in 1991 and 1994, while cases involving limiting the catching of shrimps and turtles in 1998 were hotly disputed. As forest certification started to emerge as a possible mechanism to promote more SFM in the 1990s, the WTO initially ruled that it considered mandatory certification, together with many forms of voluntary certification, as potential non-tariff barriers to trade. It took several more rounds of trade negotiations by WTO member countries to enshrine environmental and social criteria as legitimate grounds for regulating global trade (Brack, 2013).

The perceived lack of commitment to meaningful change by national governments and intergovernmental agencies resulted in the NGO community exploring alternative mechanisms to deliver more effective management and conservation of the world's forest resources.

The forestry sector was, therefore, already in a transitional phase when, in 1992, the Earth Summit was convened in Rio: it proved to be a watershed moment. The UN Conference on Environment and Development (UNCED) sought to establish a coordinated global consensus and response to the environmental and social sustainability issues that had become apparent during the previous 20 years. This included developing solutions for the sustainable exploitation of forest resources. The meeting was seen as the best opportunity to launch a new vision of sustainable development and management of the world's resources. Convened by the United Nations, the conference was hugely ambitious and wide-ranging and it generated a number of concrete proposals and initiatives covering an impressive spread of technical disciplines. These included:

- the initiation of the Convention on Biodiversity (CBD);
- the UN Framework Convention on Climate Change (UNFCCC) – which has provided the platform for much of the subsequent work on climate change and emissions;
- the UN Convention to Combat Desertification;
- the establishment of the UN Commission on Sustainable Development (CSD); and
- Agenda 21 – which provided an action plan for change.

Despite the impressive agenda and the progress made in some areas, many participants were frustrated by the debate surrounding forestry and timber issues. They deplored the failure to make any meaningful progress in the development and deployment of practical and legally binding solutions. The major forest output of the summit was a set of Principles (Forest Principles) that was intended to coordinate and guide international action on forestry: *The Non-Legally Binding Authoritative Statement of Principles for a Global Consensus on the Management, Conservation and Sustainable Development of All Types of Forests (1992).*

However, as well as being 'non-legally binding', the Principles lacked any specific targets (Wang, 2001). Already disillusioned by the lack of initial commitment shown by the ITTO, the NGO community took these principles as a signal that participating governments, and the intergovernmental bodies to which they belonged, were unwilling or unable to adopt practical and effective solutions within an acceptable timeframe; and in 1993 they announced their intention to deploy their own solution - forest certification.

3 Responses to the introduction of forest certification

The reaction to the 'launch' of forest certification was mixed. A number of key agencies, including the two principal international agencies responsible for forestry, the ITTO and the FAO, made clear their reluctance to become actively involved with the development of a viable certification scheme. The WTO, in the middle of the Uruguay round of negotiations with member countries, also made clear that they believed certification represented a threat to free trade and would ultimately prove to be a non-tariff barrier to free trade and therefore be subject to penalties.

The other international organisation which had a stake in the development of certification and standards for SFM was the International Standards Organization (ISO). Founded in 1947 and based in Switzerland, the ISO's purpose was to facilitate trade between its members by adopting standardised practices and quality. Although the ISO was not a regulatory authority it was keen to bring forest certification within its orbit and subject to its own systems of standardisation. The pressure to adopt ISO systems and processes has remained over the years and almost all of the current forest certification schemes have adopted ISO systems or processes to a greater or lesser extent. Even FSC, which at its inception specifically rejected the ISO approach as a suitable basis for its own certification scheme, has been obliged to adopt key elements of the ISO, although the desire to remain independent of the ISO membership conditions and the refusal to adopt certain ISO practices, such as mutual recognition, has meant that FSC has managed to remain outside of key ISO developments.

In general, the industry's response to the proposal to develop and introduce forest certification was negative; the industry considered certification to be fundamentally unnecessary. Other criticisms were that it would be unworkable, too complicated and expensive, and forestry and the timber trade would be unable to bear the costs: the result would be disruption to trade but with few practical benefits. It was suggested that certification would not be able to address adequately the issues relating to the management of tropical forests while the costs and economic impact would fall disproportionately on western companies and markets. Although the first part of this criticism may appear valid, the second part has proved to be patently incorrect, as the European and American suppliers have been the principal beneficiaries.

The industry was not consistently hostile; many within the sector understood that certification - of some sort - was inevitable and consequently invested in, or contributed to, developing certification systems that they could live with or which reflected their own particular requirements.

America's Sustainable Forestry Initiative (established in 1994) was developed by the American Forest and Paper Association to reflect the specific needs of this powerful industrial sector. Canada's forest industry was also proactive, and by 1996 had developed its own standard and certification systems based on an ISO platform and using standards designed for boreal forests. Both systems have subsequently been subsumed within the PEFC system.

Within the tropics, Malaysia and Indonesia responded quickly to the emergence of forest certification and sought to develop their own schemes, but development proved slow, especially for Malaysia, which struggled to develop a national approach to the standards to be used and subsequently developed a system whose rigour and objectivity prevented it from gaining international credibility or acceptance. Indonesia had a significant advantage through its partnership with the Rainforest Alliance and as early as 1990 had already established its own scheme, Lembaga Ekolabel Indonesia (LEI). This was technically quite sophisticated, but struggled to gain broad international acceptance in the absence of a clear demand for its certificates.

4 The development of forest certification

4.1 Forest Stewardship Council (FSC) certification

In 1993 the most tangible and significant developments were being driven by NGOs. The NGO announcement to develop forest certification as a means of addressing weak control and failure to achieve sustainability of the world's forest sources may have been borne out of frustration, but it was not made without preparation.

Since many agencies, including the ITTO and the WTO, had expressed strong reservations about certification, in 1991, the WWF, together with a number of other NGOs, formed a forest certification working group. Its aim was to develop a practical, effective certification system that was appropriate for the tropics and which would ensure the long-term sustainability of forest resources. This initial focus on tropical forests had to be dropped in favour of a scheme that could be applied globally, to all forest types and irrespective of geographical location; this was essential to avoid certification being rejected as a non-tariff barrier to trade. Considerable development work had already been undertaken over the previous two years and the working group was able to build on this. Particularly notable is the work of the Rainforest Alliance, which had developed its prototype Smartwood certification system while working with Indonesia's national forest company in the late 1980s: the Rainforest Alliance was a key member of the certification working group.

The NGO declaration of 1993 was followed quickly by the formal establishment of the first dedicated global forest certification scheme; this was the FSC (https://fsc.org/en). The FSC's founding assembly was held in Canada in October 1993. The FSC became a legal entity in 1994 with its headquarters in Mexico, but it still took a further two years before a fully operational certification system would be available to commercial customers. This arrived in early 1996 with the FSC's accreditation of four certification bodies (CBs), which were authorised to conduct certification and issue certificates based on FSC certification standards.

The working group, and its NGO backers that developed the FSC, had to develop an organisational structure and operational framework that was unlike anything that had come before. Certification had already become a recognised tool for quality management under the ISO and its ISO 9000 series, but the NGO's vision for its forest management scheme was far more ambitious. The aspiration was to establish a certification scheme that not only ensured that its certificate holders met pre-determined levels of forest management, but offered a credible pathway for companies to change their standard operating practices so as to meet the FSC's forest standards. This required the scheme to offer some form of incentive that would be attractive to commercial companies.

The working group's design brief was to construct an organisational structure, together with a form of governance, that reflected the principles of transparency, inclusiveness and sustainability to which the stakeholders were committed and which they were expecting from the companies that gained their certificate. The organisation needed to be independent, free from control by any form of national governmental or inter-governmental body, but at the same time it had to respect national and international law. Finally, the FSC needed to accommodate the practical and commercial realities of the forest and timber trades that they were seeking to engage and change.

The design parameters were complex and demanding. Although the stakeholder support base was primarily derived from NGOs, this still represented a wide range of different, and sometimes contradictory, aspirations and expectations. The inevitable result was that the designers of the FSC were obliged to compromise, adopting mechanisms or design features that sought to meet the majority of stakeholders' aspirations but sometimes by adopting a less than ideal solution. As an example, resistance from a proportion of its stakeholder base led to the FSC abandoning its original preference for establishing the FSC as a foundation and adopting, instead, a governance structure based on membership. The FSC was established with two chambers representing environmental and social interests, which was soon expanded to three chambers with the inclusion of an economic chamber. This structure contributed to the FSC's reputation for a representative governance structure of admirable inclusiveness and transparency, but it could also be considered a less than ideal structure for responding quickly to changing circumstances.

Crucially, the original working group included representatives from the commercial timber trade, including one of the largest retail chains selling timber and wood products that had been sourced from all over the world. In the late 1980s and early 1990s, the UK retailer B&Q, along with many others in the international timber trade, had been on the receiving end of direct action by environmental NGOs. This included timber boycotts, protests and negative press campaigns such as the 'Mahogany is Murder' campaign (Doyle and MacGregor, 2013).

In response, B&Q requested their suppliers to identify the sources of their timber. The results so alarmed senior managers that B&Q made a corporate commitment to purchase only sustainably managed timber and, as part of this strategy, it became an active member of the certification working group. The importance of the involvement of B&Q, together with the other processors and retailers who made an early commitment to purchase FSC-certified product, cannot be underestimated and it was to prove a key factor in the rapid take-up of FSC-certification once it had been launched.

The designers of the FSC had learned the lessons from the development of organic farming, a movement that was seeking to develop a more sustainable form of agriculture in contrast to intensive industrial farming methods. Established in the United Kingdom in 1947, the founding organisation – the Soil Association – had developed a unique and innovative prototype of certification for organically grown agricultural products. However, the uptake of organic farming practices had progressed slowly and by the 1990s it was estimated that less than 1% of UK agriculture had 'gone organic'. The majority of farmers remained unconvinced of the benefits and mainstream agriculture remained resistant to the wider adoption of organic practices. While the movement was respected for its ethical stand and its advocacy skills, it was clear that it had failed to harness the purchasing power of the general public, which was potentially the most effective means of inducing producers to change their practices.

Those responsible for developing forest certification realised a market for FSC-certified products would need to exist prior to the launch of the scheme. From 1993, the WWF coordinated a range of activities to establish this demand, not only by coordinating efforts to develop appropriate purchasing strategies by the commercial companies who were part of the working group, but also by inducing other companies to commit to preferable purchasing of FSC-certified products. As a result, the FSC was strongly criticised for unfair market manipulation, with the most strident complaints coming from timber-producing countries and organisations that resented being 'pressured' into meeting the new requirements demanded by their customers. Nonetheless, FSC-certified material became available through wholesale and retail outlets very quickly, and in reasonable volumes, as soon as the scheme became operational in early 1996, although the species and product ranges covered were limited.

The benefits of having a committed and active commercial stakeholder group within the FSC membership also had its disadvantages. Accommodating the legitimate concerns and requirements of its commercial and economic stakeholders was particularly challenging for the FSC and had been met, in part, by the adoption of the third chamber of governance, the economic chamber. Commercial organisations, however committed they might be to the principle of sustainable management of its resource base, still require that standards and operational practices be sufficiently simple and standardised to be implemented cost-effectively, thus potentially making them less rigorous.

The FSC also faced a more fundamental and insidious threat, and one which was anticipated even during its development: once it became operational, forest certification was effectively a 'commercial business', albeit one offering a sustainability service and its success or failure would be determined by the demand for its products and services and the requirements of its customer base. This raised the risk that compromises might have to be made in order to keep customers as part of the scheme.

One solution to this potential 'conflict of interest' was to use a central fund - raised from a levy on all timber transactions - which would meet the audit costs of all applicant companies. The certification body (CB) could be selected from a panel of accredited CBs through an objective selection process and the scheme owner (FSC) would be responsible for the issuing of the certificate based on the audit report submitted to them by the CB, thus separating the CB from any direct commercial or financial link with the company it was certifying. Unfortunately, this novel structure proved to be too demanding to incorporate.

The designers of the FSC were also constrained more directly by external pressures and circumstances that effectively prohibited certain strategic options or dictated the inclusion of specific design characteristics. The most obvious of these was the need to comply with the trading regulations set by the WTO. These required the FSC to ensure that, whatever design was finally adopted, the certification scheme would not be considered as a non-tariff constraint of trade. Furthermore, certification could not be made mandatory - it had to be a voluntary exercise, freely entered into - and, consequently, great care had to be exercised in the design of the FSC to ensure that any requirements that related to legal compliance (national or international) were framed appropriately.

At the heart of any certification scheme are standards and the FSC had developed its own integrated set of principles, criteria and indicators which they believed described good forest management and responsible operational practice. The hierarchical framework of Principles, Criteria and Indicators had become the most widely adopted format for standards and were commonly abbreviated as P & C or C & I (Lammerts von Bueren and Blom, 1997). The FSC P & C represented a broad range of environmental and social principles and outcomes that defined good forest management. These P & C fulfilled three related functions:

- they represented the core principles of the FSC and its members;
- they provided the fundamental framework around which FSC's national initiatives could construct national, local or regional standards while ensuring that they all remained compatible; and
- with the addition of appropriate indicators and verifiers, they became the FSC certification standards.

In addition, the FSC P & C included novel features such as principles and criteria related to Indigenous People's Rights and Community Relations.

In early 1996, the FSC became the first practical, international forest certification scheme that was available to commercial companies. It is important to understand that the certification model that had been developed was not simply conventional certification aimed at the forest industry. What the FSC certification was offering to mainstream timber producers and processors was a process that used certification as the medium, and the driver, for ensuring forest producers and processors adhered to basic standards of good forest practice. The FSC certification scheme was designed to ensure minimum performance thresholds were met across a range of environmental and social criteria.

Even more radical, and controversial, was that those organisations which met the standards - and gained the FSC certificate - would be rewarded by access to specific markets: markets that demanded certification and excluded non-certified timber or wood products, even if they met the same quality specifications as certified timber (Bass, 1996). The FSC's specific form of certification for forests (and timber and wood products) also included a unique and critical fourth element: the FSC certification linked the certificate with the end product (the timber or finished product) throughout the entire length of the supply chain - what is now called the 'chain of custody'. This provided the evidence required to assure end-users, and the general public, that the product had actually been sourced from the producer organisation that had received the certificate.

The FSC resisted considerable pressure to adopt a system that was compatible with the standard setting and auditing systems adopted by the ISO. This independence conveyed a number of advantages, including the freedom to consider, or refuse, demands for mutual recognition from other certification schemes. But the FSC's independence from the ISO process meant that it remained outside of the international mainstream of standard setting, as most countries were signatories to the ISO and developed their standards within the national membership structure of the ISO.

This represented a significant disadvantage when the FSC was seeking to influence the international debate on standards, audit processes, accreditation and the acceptability of certification schemes in the international market place. Over the last 20 years, the FSC has progressively adopted many of the elements of the ISO process and incorporated a number of ISO standards, particularly those relating to accreditation, audit process and auditor competence and the FSC now has standards and processes that are ISO-compatible; but this has come at a cost, with many of the original stakeholders expressing concerns that adopting a more ISO-friendly system has compromised the FSC's ability to adopt individual, rigorous or forest-specific or location-specific solutions. (Bass, 1998; McDermott, 2011)

The FSC, like any organisation, has had to respond to new demands and a changing operating environment. These changes have been required to meet the demands of the market and of its own members, supporters, certificate holders and critics. Very early in its development, the FSC was forced to abandon its emphasis on tropical forest sources, and re-design its systems and standards to ensure that its certifications were genuinely global in scope and able to accommodate any type of forest, both natural and planted. The issue of planted forests – and the associated issue of forest conversion – have together proved to be critical issues that continue to be a source of disagreement within the FSC community and which the FSC has struggled to resolve to the satisfaction of its members, supporters and certificate holders. In February 2020, the FSC issued another Consultation Paper in an effort to find a practical and workable solution to the issue of forest conversion. The FSC has also developed new technical standards for specific situations: these include plantations; multiple sites; non-timber forest crops; and even non-certified sources of timber used by its certificate holders as part of composite wood-based products.

The FSC has been forced to respond to commercial pressures, not only those affecting the supply and movement of timber and forest products (such as the Lacey Act in the United States and the European Union Timber Regulation), but also those pressures and international regulations that relate to certification and accreditation bodies. It has been forced over time to change its governance structures and the systems of oversight and accreditation by which it regulates the activities of its CBs. It has also had to respond to the practical difficulties of running a commercial business, including: seeking appropriate sources of funding and investment; developing practical alternative sources of income and cash flow; and new systems for evaluation and monitoring of the performance of the scheme and its certificate holders.

These changes, such as the inclusion (and retention) of plantations and particularly the inclusion of logging in primary natural forests within the scope of the certification scheme, the devolution of power and executive authority to CBs, and the increasing inclusion of ISO systems and processes, have caused tensions within the FSC membership and its supporter base. The result has been a slow erosion of support from NGOs and the withdrawal of certain key stakeholders and stakeholder groups, some of which have subsequently formed their own organisations or actively promoted their own solutions (Changing Markets Foundation, 2018). The complexity of the arguments related to key issues is demonstrated by the protracted debate concerning the protection of Intact Forest Landscapes and Indigenous Cultural Landscapes, which started in FSC in 2014 with Motion 65 and is still ongoing in 2020 (https://fsc.org/en/p age/intact-forest-landscapes).

Of far more concern is the increasing body of evidence of flaws within the FSC's certification programme, whether from former FSC supporters or from 'watchdog' organisations independently monitoring industries which have demonstrated a poor environmental or social track record. These have highlighted examples where certified products have been found not to have been sourced responsibly, undermining the credibility of the certification process. Examples include: FSC-Watch (https://fsc-watch.com/); Forest Monitor (http://www.forest-monitor.com/en/); Environmental Investigation Agency (https://eia-international.org/); Finnwatch (https://finnwatch.org/en/); and Global Witness (https://www w.globalwitness.org/en/campaigns/forests/independent-forest-monitoring-ifm/).

What have been seen as unsatisfactory responses by FSC's managers and board, and the perception of a flawed system for dealing with and resolving complaints, have led to the progressive withdrawal of support by key NGOs including: Swedish Society for Nature Conservation in 2008, Friends of the Earth in 2009, Tropical Forest Trust in 2013 and Greenpeace in 2018. In 2014 a previous CEO of FSC raised concerns about the effectiveness and level of assurance of FSC's chain of custody system, echoing similar comments by the Tropical Forest Trust when they withdrew their support. These criticisms have called into question the credibility of the entire forest certification industry, not just certification by FSC. It is important to note that other NGOs, including WWF, have been more positive in their assessment of the impact of the FSC (WWF, 2014). This echoes other research on the beneficial impacts of FSC certification (Burivalova et al., 2017).

4.2 The Programme for the Endorsement of Forest Certification (PEFC)

The Programme for the Endorsement of Forest Certification (PEFC) scheme is the only other forest certification scheme that can be called truly international in scope although it adopts a different approach to that of other certification schemes (https://www.pefc.org/). The PEFC scheme developed from a schism within NGO movements seeking to establish a viable certification scheme after the 1993 declaration. Originally conceived as the Pan European

Forest Certification scheme, the PEFC was initially developed by European stakeholders who had become disillusioned by the FSC's focus on tropical countries and what they perceived as the FSC's failure to represent adequately small landowners in Scandinavia.

The PEFC was founded in 1999 in response to the specific requirements of small, family-owned and community forests by 11 associations from European countries representing these groups. The initiative attempted to overcome some of the difficulties faced by smallholders and community groups in complying with FSC certification requirements and has subsequently evolved into a scheme which in 2004 became the largest scheme in the world, by certified area (Rametsteiner et al., 1999; Thornber et al., 1999).

The PEFC was based on a number of intergovernmental processes (Ministerial Conference for the Protection of Forest in Europe, the Montreal Process and the ATO/ITTO process). It requires that all national members (the organisations responsible for the implementation of PEFC certification within their countries) have the support of the relevant national forest owners or national government forestry organisation. Forest owners have always been represented on the Board of Directors at PEFC International, together with NGOs, labour organisations, researchers, companies and other key stakeholders. Forest owners and their knowledge and expertise play a central role in managing and developing sustainable forestry. The competence and long-term commitment of forest owners are seen as useful drivers in the efforts to make forestry more sustainable.

The PEFC utilised a different format for its standards (initially basing them on the Helsinki P & C which had not been designed as auditable standards) and contained no chain of custody component. It offered an alternative approach based on the development of national or local standards through a nationally acceptable process, and was broadly ISO-compatible. The multiple standards developed would be the basis of national certification conducted by commercial CBs, but the standards would be 'harmonised' and mutually recognised through a process overseen and authorised by the PEFC.

National standard development is undertaken by multi-stakeholder working groups, with the composition based on nine major stakeholder groups as defined by Agenda 21 of the United Nations Conference on Environment and Development (UNCED) in Rio de Janeiro in 1992. Standards must be developed by the national standard setting body in a process that is open, transparent and based on consensus among a broad range of stakeholders (ISO/IEC Guide 59 and Guide 2). The PEFC requires that forest owners are always included as stakeholders in the development of forest management standards, given that they are affected by and influence the implementation of the standard. As such, forest owners must be represented in an appropriate share among participants (PEFC ST 1001:2017 6.4.2).

It is worth noting that standards can only be approved by the working group on the basis of consensus by all stakeholders, including forest owners (PEFC ST 1001:20176 6.4.5). This process is similar to that of the FSC process, where national standards are developed by multi-stakeholder groups (National Initiatives) and adopting a process that meets the requirements and standards laid down by the FSC (FSC Standard FSC-STD-60-002 v1-0 and procedure FSC-PRO-60-007 V1-2) and which is endorsed by the FSC governing body. The PEFC has established a bottom-up approach, requiring all standards to be developed at national level in compliance with the PEFC's globally applicable sustainability benchmarks.

The PEFC scheme exists and reports as a single scheme but it is comprised of multiple standards, each established by an appropriate national process, which attempt to define criteria and indicators that reflect the specific needs of its stakeholders. The PEFC then harmonises these standards, ensuring their compatibility and congruence with other national standards, so that they can be mutually recognised by other PEFC members.

This methodology aims to ensure consideration of the local dimensions of sustainability as well as compliance with the existing forest governance regime and regulatory framework. The latest version of the PEFC forest management standard has been adapted to follow the structure of ISO standards and processes and strictly separates standard setting, certification and accreditation, building on internationally recognised requirements for certification and accreditation defined by ISO and the International Accreditation Forum (IAF) and its certification is carried out by impartial, independent third party auditing companies. PEFC certification, as with FSC, is consequently delivered through these third parties and both suffer the same inherent weaknesses: they are paid directly by the client company and all relationships between the certification scheme and certificate holder are mediated through this third party.

The relationship – the responsibilities and obligations – between the scheme owner (PEFC) and its CBs is maintained through accreditation contracts. For the PEFC, the accreditation process is determined by the International Accreditation Forum (IAF) and the IAF's Regional Accreditation Groups and Multilateral Recognition Arrangement (MLA). The MLA ensures that accreditation programmes are operated consistently and in an equivalent way, based on peer evaluation (PEFC ST 2003:2012 Second Edition; Annex 6). As mentioned previously, this is different from the FSC scheme, which is not a member of the IAF. The FSC's accreditation procedures follow a very similar pattern and adhere to ISO guidelines but oversight is provided by ISEAL Alliance, which specialises in maintenance and oversight of sustainability certification schemes (http://www.isealalliance.org/).

Subsequently, the PEFC has expanded its methodology of mutual recognition more widely through the restructured organisation, which was renamed the

Programme for the Endorsement of Forest Certification Schemes in 2003. This offered a standard setting mechanism that had global scope, endorsing and mutually recognising standards from all over the world, and this contributed to the rapid expansion of the PEFC and overtaking the FSC (in terms of certified area) in 2004. It continues to be the scheme with the largest certified area.

The equivalence of PEFC and FSC certification in terms of meeting European definitions of legality and sustainability means that both are considered as acceptable in terms of legality and sustainability in spite of considerable differences in structure and standards. Many companies possess 'double' certification (FSC and PEFC) to ensure they meet all the requirements of key markets (https://www.pefc.org/news/double-certification-on-the-rise-joint-pefc-fsc-data-shows).

Within the PEFC, new national standards and certification schemes are being developed rapidly and the endorsement of schemes generates rapid increases in certified area. Recently endorsed standards and schemes include PEFC Annual Report (2019) and India, Bulgaria, Thailand and Cameroon in 2019, while other countries are actively seeking membership and developing their own standards and schemes, including Vietnam.

4.3 Other certification schemes

The detailed explanation of the history and development of FSC and PEFC certification schemes is not intended to reflect any inherent superiority or any personal preference on the part of the author. It simply reflects the current market preferences and the domination of the certification market by these two schemes. The FSC was the first viable forest certification scheme that was available to producers, exporters, processors, manufacturers and retailers of timber and forest products globally. The FSC's developers were able to draw on the pioneering work of a number of its stakeholders, particularly the work of the Rainforest Alliance whose prototype certification scheme in Indonesia in the early 1980s has already been mentioned and which enabled FSC to anticipate or overcome many of the early problems facing the certification of forests.

Once launched, the FSC became the first commercial certification scheme to encounter the wide range of ethical, operational and commercial problems which were unique to certifiers of forests and the timber trade worldwide. The FSC was also the first scheme to develop practical and cost-effective solutions. Those organisations whose establishment succeeded the FSC were able to benefit from this pioneering work, helped by the FSC's policy of transparency, which permitted many of these practical solutions – and the collaborative process through which they had been formulated – to be made widely available.

Many of the other forest certification schemes owe their existence, directly or indirectly, to the presence of the FSC, or to the original declaration by

NGOs in 1993 to develop a viable model of forest certification. Some schemes developed out of the conflicting aspirations of different stakeholder groups, either within the original certification working group (from which the FSC evolved) or from within the FSC's own members and supporters. Such was the case with the PEFC which overcame the FSC's early advantages, evolved and finally surpassed the FSC in terms of certified forest area.

Indonesia was the first country to develop its own national forest certification scheme when it launched its Lembaga Ekolabel Indonesia (LEI) in 1993 (Elliott, 2000; Muhtaman and Prasetyo, 2006). Although LEI predated the launch of the better-known FSC scheme, it struggled to gain a sufficient volume of commercial clients or international acceptance, in spite of considerable support from a number of key environmental NGOs. LEI tried to build on the momentum generated by the formation of the FSC in 1994 and sought mutual recognition with the FSC. FSC members were opposed in principle to any form of mutual recognition but a form of 'joint certification' – using audits conducted against LEI's Criteria and Indicators together with the FSC's P & C – was authorised under a Joint Certification Protocol agreed in 1998.

But forestry in Indonesia suffered from a number of fundamental and chronic problems; exacerbated by weak regulation and control, these proved to be issues of sustainability that certification, on its own, was poorly equipped to resolve. Additionally, LEI was established in the absence of any clear commercial demand, or commitment from customers, for LEI-certified timber products. Indonesia developed a complex array of different measures in an attempt to resolve some of the key issues, but the inability to address forest-related conflicts related to indigenous peoples' land rights has proved to be a major obstacle that has hampered LEI and the credibility of its certification.

While many challenges remain, a few positive impacts of LEI certification have been noted: these include the establishment of a government incentive for companies to pass LEI certification; an increased willingness of companies to engage in public consultation; and the opening up of political space for NGOs and communities to express their concerns (Muhtaman and Prasetyo, 2006). Certification in Indonesia has changed direction with the focus now firmly on legality and this has led to Indonesia being the first country to develop a legality licensing scheme under the EU FLEGT Voluntary Partnership Agreement.

Some certification schemes were conceived in order to satisfy specific local requirements or to better meet particular stakeholder groups' expectations and the North American and Canadian SFI and CSA schemes have already been mentioned. After some years of offering an alternative form of 'independent' certification, these schemes sought to benefit from the equivalence of PEFC and FSC schemes in Europe, and sought endorsement under PEFC's mutual recognition: from 2004 to 2005 both these schemes have been subsumed within the PEFC scheme.

Other national schemes have developed with varying degrees of success. China developed its own certification system from 2001 – partly to address concerns related to issues of sovereignty – and the China Forest Certification Scheme became fully operational in 2010. In 2014 it was endorsed by the PEFC. In contrast Ghana has made at least two attempts to develop its own system without conspicuous success and has finally chosen to develop a FLEGT legality licensing system.

There was also a period which saw a drive for regional certification schemes and these were pursued in ASEAN countries, Melanesia and Francophone African countries. With the expansion of PEFC the appetite for such schemes has waned. Additionally, the large schemes have the technical capacity and experience to develop specific sub-programmes to deal with local or regional issues. Small and low-intensity management, community forestry and forest/ agriculture combinations are types of forest management for which FSC and PEFC have developed specific solutions.

The other scheme which needs to be mentioned is the ISO 14000 system and its current version 14001. Following the lead of the standard for environmental control released by British Standards (BS 7750) in 1992, ISO developed its own series of international environmental management standards in 1996. Like many ISO standards there were no specific performance thresholds, although the ISO 14001 system was immeasurably improved with an obligation to adhere to the national environmental laws.

ISO 14000 does provide an auditing framework but its emphasis on the development of systems, rather than specific performance thresholds, means that it has not been considered an effective vehicle for ensuring sustainability for the forestry and wood processing industries and it has not been widely or consistently adopted. The ISO appears to be used, principally, as a bridging or enabling mechanism to the adoption of one of the recognised forest certification schemes. ISO 14000 series certification can be considered as the closest representation of 'standard certification'. An analysis of ISO 14001 systems, conducted in 2004, provides a useful summary of the key issues (Neumayer and Perkins, 2004).

The certification industry has become much simpler with the endorsement of many of the original independent schemes by the PEFC.

5 Rates and spread of certification

Certification was designed to support SFM in tropical production forests. While it can therefore assist in control of forest loss in these forests, it cannot in itself address all the other drivers of deforestation such as expanding commercial and subsistence agriculture, rural poverty, urbanisation and infrastructure development or events such as wildfires, which are exacerbated by improved access to forests. Certification can only take place when the forest owner and/ or lessee commit to the process; without appropriate commitment certification alone will not necessarily prevent forest loss and reduce levels of degradation. Nevertheless, areas of forests certified does provide an indication of the proportion of forest owners and occupiers who are interested in attaining some form of sustainable forest management.

Figure 1 shows the growth in forest area covered by certification schemes until 2016. The total area of certified forest has increased further since 2016. By mid-2019 the total area of forests certified is reported as 325 million ha for the PEFC and a little over 200 million ha for the FSC. As this includes approximately 92 million ha of forest that has been certified under both schemes, the figures need to be treated with care (see Table 1).

Interestingly, in spite of the different motivations, aspirations and development pathways followed by the two main certification schemes, the FSC demonstrates the same broad geographic imbalance in the locations of its certified areas (see Table 1). Although schemes do not maintain identical regional groupings of countries, they are essentially very similar. This table shows over 50% of the FSC's certified areas fall in Europe. In spite of containing major timber producers such as Malaysia, Indonesia, Papua New Guinea, and Asia accounts for only 5% of the FSC's certified area, with Oceania (including Australia and New Zealand) even less at 1.3%. Africa is also poorly represented with only 3%. A more accurate total is 430 million ha. This is a large area of

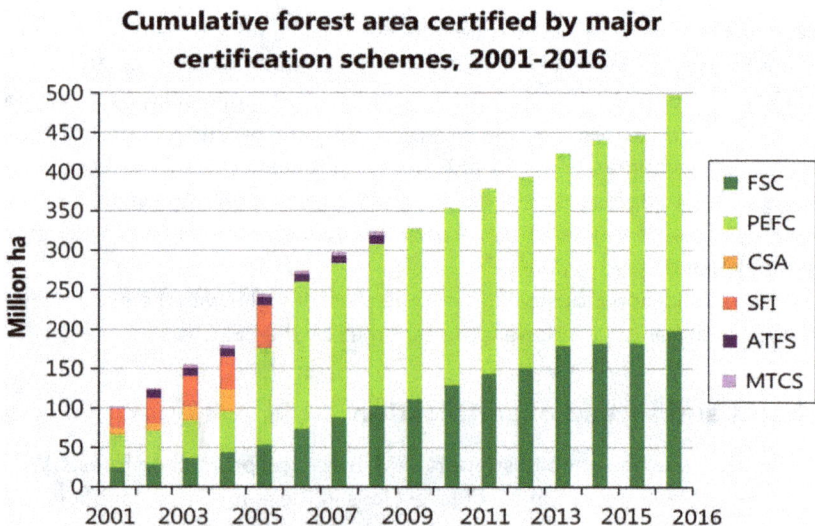

Figure 1 Cumulative forest area certified by major forest certification schemes 2001-2016. (Source: UK Timber Trades Federation: https://timbertradefederation.files.word press.com/2017/08/slide17.png).

Table 1 Comparative breakdown of certified forest areas by geographic region

Region	PEFC % of total certified area 2019[a]	FSC % of total certified area 2019[a]
Europe	36	51
Africa	<1	3
North America	53	33
Central and South America	3	7
Asia	5	5
Oceania	3	1

[a] Note: The total certified area reported by the PEFC in 2019 is 325 million ha and by the FSC is 211 million ha. However, both totals include areas that have been certified by both schemes. The double-certified area is 93 million ha, and so the total area certified is 325 + 211 − 93 = 443 million ha.

forest and represents a significant success in terms of the uptake of commercial forest certification within the forest and timber industry. It is estimated that this forest area represents 10.7% of a total global forest area of 4.03 billion ha.

In terms of productive forests, the certified forests of the FSC and PEFC schemes are the source of approximately 29.6% of all the roundwood traded annually throughout the world (Yale School of Forestry and Environmental Studies: Global Forest Atlas: https://globalforestatlas.yale.edu/conservation/forest-certification). These figures show the importance of certified forests to the world's economy and timber trade, but they provide no direct evidence about the contribution of certification to the sustainable management of those forests.

Given that forest certification was designed to support SFM in tropical production forests, it is disappointing that it has been applied predominantly in the temperate forests of Europe and North America; there are many reasons for this. The first is that FSC certification was initially fiercely resisted by many of the largest owners and managers of tropical forests and suppliers of tropical timber as discriminatory and a potential constraint to free trade. The FSC was obliged to become a global scheme in response to this criticism. Early progress in certified forests, however, was made not in the tropics (with a few notable exceptions) but in temperate and boreal areas where forests were generally already better regulated and better managed; certification was easier to achieve in these forests.

Where the objectives of forest management involve conservation, protection and maintenance of a forest area, certification has proved to be a less valuable tool than many anticipated, mainly because the costs for certification are considered as very high. The complexity and cost of independent third-party certification, and the ongoing costs of maintaining the certificate, are significant and can only be justified if certification results in a commensurate income stream to offset these costs. Precious Woods has had its forests in the Amazon and Gabon re-certified several times; it has been able to do this because of the commitment of its stakeholders and because the markets have been willing to pay a premium for the timber from these certified concessions.

For forests which are predicated on minimal levels of exploitation, certification is clearly not a financially viable solution. Although it offers an effective means of demonstrating sustainability, the costs of certification may simply exceed the practical benefits and owners and managers may prefer, or be constrained to adopt, cheaper or more cost-effective solutions. Forest certification has also been utilised as a means of attracting green investors and providing an independent assurance of a forestry company's commitment to the principles of good management which are environmentally and socially responsible: in this case the cost of certification is justified by the possibility of capturing appropriate levels of investment from committed investors, such as the Triodos Bank.

Certification has not been able to offer a meaningful solution where those with authority for the forests have no intention of managing them responsibly, sustainably or even legally. Regrettably, large areas of the world's forests fall into this category. This has contributed to the poor track record of SFM in Africa and Asia and, consequently, the poor uptake of forest certification (Blaser et al., 2011). Figure 2 shows the leading areas of deforestation that have changed over the last 13 years but are still predominantly from the tropical regions. One of the most consistent criticisms levelled at forest certification is that it has signally failed to make progress in ensuring or promoting SFM in these most critical regions.

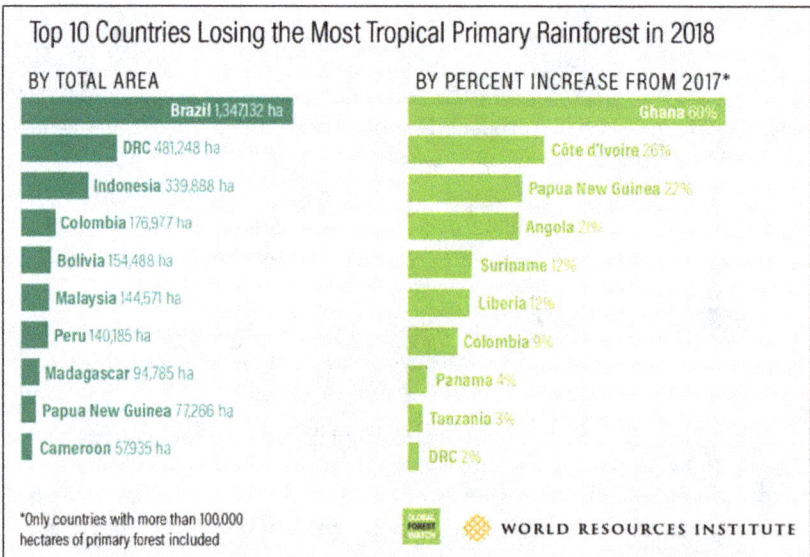

Figure 2 Top 10 countries losing tropical rainforest in 2018. (Source: http://www.migh tyearth.org/heres-what-deforestation-looks-like-in-2019-and-what-we-can-do-about-it/).

6 Assessing the impact of forest certification

The previous discussion has shown the complexity of the current forest certification 'industry' and the differing aspirations and motivations that led to their establishment. Although sustainability, good forest management and sensitivity to environmental and social issues are goals enshrined in the mission statements of all forest certification schemes, they are defined differently, assigned different levels of priority and may even be pursued with different levels of commitment and rigour.

All certification schemes are comprised of four key components:

- Standards – which define the minimum acceptable level that must be achieved;
- Inspection/audit – the process that gathers the evidence of an organisation's performance against the standard;
- Certification decision – the process where the results of the audit are analysed and a decision made as to the level of compliance of the organisation with the standard; and
- Certification, Licencing and Labelling – the process by which the certificate is issued (or refused), together with the terms and conditions of being a certificate holder, the conditions required for maintaining certification, and any rights, benefits or obligations required of the company while a certificate holder.

Certificate holders, the customers for certified products and the wider general public rely on this certification process to deliver on its goal of SFM, and will only place their trust in the process (and the certificate it generates) if they are confident that the certification process has been conducted objectively, competently and with appropriate levels of rigour and commitment.

Currently, there remain concerns that not all forest certification schemes address these four components with equal levels of objectivity and rigour and a number of stakeholder groups remain unconvinced that any of the current models of forest certification can deliver all their intended benefits (Jakub, 2019).

One contentious area has been defining the scope of certification. FSC stakeholders in particular have been particularly concerned with the high proportion of plantation forests within the FSC certified supply base. The original drive for developing certification was to address the decline of natural forests, and many stakeholders wanted to exclude plantations altogether from the scope of FSC certification.

Initially, the FSC's 9 Principles attempted to address all forests and did not distinguish plantations until 1996 when the FSC board introduced Principle 10, which defined criteria and indicators specifically framed for forest plantations. The FSC as an organisation has accepted that plantations are an integral component of the forestry sector: in many countries they form a significant part of the country's resource base and revenue-generating potential. However, the FSC has repeatedly had to defend the maintenance of plantations within the scope of the scheme and has reviewed the status of plantations and related issues in 2004, 2011, 2014 and more recently in 2019.

A closely related issue within the FSC's certification approach is the issue of 'conversion' from one type of forest to another, for example, from an 'unmanaged' natural forest to a 'managed' natural forest, applying sustained yield management and silvicultural practices. Similar problems exist for conversion from forest to plantation; plantation to plantation of a different species or mix of species; and even from non-forest to forest. Under what conditions a certification scheme can technically, and philosophically, accept the conversion of one forest type to another and still be considered to meet the definition of SFM is one of many fundamental questions that continue to challenge forest certification schemes.

In 2011 approximately 16% of the FSC's certified forest area was forest plantation. In 2017 the area of plantation increased to 17 million ha in absolute terms, but in relative terms the certified planted forest area was only 9%. This compares to 51 million ha of 'semi-natural, mixed plantations and natural forests', representing 65% of the total. Further changes to the FSC's certification systems have meant that this issue has resulted in some strange outcomes. As an example, timber generated by the conversion of low productivity forest plantations to oil palm cultivation has been included in the category 'controlled wood', a permissible source that can be mixed with FSC timber and subsequently labelled as an FSC product. Forest plantations of exotic species in China have been certified even though they have been established through the eradication of previously degraded natural forest.

The PEFC, and the schemes it has endorsed, have generally accommodated plantations within their certification more easily than the FSC, but the result is that plantations and semi-natural forests represent a substantial proportion of the area of forests certified by both schemes. Both schemes have had to continue to refine their standards to deal with the issues of plantations and conversion and have also introduced retrospective measures to deal with previous examples of conversion. These practices have generated considerable criticism (especially for the FSC). The issue of conversion is one of the biggest differences between the schemes and has a key impact on the credibility of both schemes.

So far, the FSC has struggled to develop a solution that sits comfortably with the bulk of its membership and supporters. This is the result of the high bar that was set initially by its NGO members who are unwilling to compromise on a matter of principle, as evidenced by the failure of successive review processes to propose a satisfactory solution. The arguments related to the validity of plantations may seem arcane, but the related issue of conversion has proved to be a critical issue of sustainability and one of the critical pivots around which the effectiveness and credibility of certification revolve.

Perhaps the most contentious area has been the failure of the certification process to deliver sustainable outcomes in practice. Most certification schemes have adopted the internationally-recognised system of using independent, third-party CBs to conduct audits against the standards within a particular scheme. The schemes devolve specific responsibilities, authorise a range of activities and control the operations of their CBs (many of whom work for multiple certifications schemes) through accreditation agreements and contracts. These determine not only the terms of engagement between the scheme owner and the CB, but also between the CB and the client (certificate holder).

This approach contains a number of potential weaknesses and maintaining this relationship requires constant oversight. Eliminating the 'direct' relationship between scheme owner and its certificate holder means that any monitoring and control of certificate holders can only be conducted through a third party, the CB, which is also responsible for conducting the audit process and making the certification decision. This makes process of oversight by owners of schemes less direct and more complicated, requiring the cooperation and agreement of two organisations rather than one. Generally, it is more time-consuming, as well as being more prone to misunderstanding and confusion. ISO has developed a structured process that scheme owners can adopt to manage this relationship, and most forest certification schemes have adopted this process. However, many stakeholders (particularly those involved in the FSC) feel that forest certification has been weakened by adopting a generic solution, which has been designed to suit conventional business conditions.

A number of NGO and civil society organisations continue to feel that the closer forest certification approaches 'conventional certification' (as structured by ISO), the less effective it will be at delivering SFM; this has certainly been a contributory factor in the FSC's loss of NGO support over the past decade.

NGO support has waned due to the apparent increase in the number and severity of complaints and disputes relating to the performance of certified companies, and the accumulation of evidence about significant systematic failures in high-profile certifications. FSC-Watch (www.fsc-watch.com), which focusses exclusively on infractions identified in FSC certifications, presents a representative selection of these infractions. These demonstrate a succession of high-profile 'failures' of FSC certification in 15 different countries over the past 20 years, certified by a number of different CBs. But the failures are not unique to the FSC. PEFC certification is subject to less scrutiny, partly because the NGOs and CSOs that make up the bulk of the watchdog organisations have already made clear that they believe the PEFC scheme's standards and approach represent a fundamentally less rigorous approach to certification, but failures have been identified within PEFC-certified forests as well.

The concern is that the abuses and failures are continuing to occur within certified forests, casting doubt on the effectiveness of the certification process. It is perhaps not surprising that this is the case. Once certification schemes had exhausted the 'low hanging fruit' they were obliged to consider certifying organisations whose management and practises were considered to represent 'less sustainable' models of forest management. The motivations and commitment of a number of these organizations were more variable and many had been induced to undertake certification by the demands of their customers or pressures applied by investors or stakeholders, but without sufficient preparation.

CBs were increasingly required to conduct certifications that were fundamentally more challenging and certification results were correspondingly more equivocal. Correcting non-compliances was likely to require more fundamental changes of management and operational practice and tended to be more susceptible to subsequent reversion to previous practice unless constantly monitored. Furthermore, these failures are more easily spotted than before, given the greater level of transparency and the documentary evidence found in forests areas under management that have been independently audited and certified.

One simple conclusion that can be drawn is that certification may reduce but does not eliminate entirely the chance of failures, and the potential for failure is commensurately higher in organisations whose commitment to sustainable practises is driven by commercial or political considerations. Changes of circumstance are, therefore, more likely to result in a return to previous practice. Such changes include a drop in prices for timber or forest products; increased management complexity and demands on staff from compliance; and additional costs due to the need to adopt safer, more sustainable working practices. Failures and non-compliance should be identified by CBs during their annual audit; however, having conducted a costly and lengthy main assessment, sometimes over many weeks, and issued a certificate for a fixed period, CBs are reluctant to withdraw the certificate for a single infraction identified during an annual visit. They prefer to resolve the issue through corrective actions which they can monitor during the period of the certificate.

Unfortunately, unless carefully and rigorously monitored, this can permit chronic, long-term failures to become a regularly occurring feature. It is also the case that the level of commercial and market benefits generated by certification is frequently insufficient to induce the operators to change fundamentally the way they operate. It is easy to undertake corrective actions which are cosmetic but which do not resolve the real problem; more worryingly, some may even be deliberately designed to avoid resolving the problem. This is in spite of the adoption of increasingly sophisticated forms of root cause analysis and integrated corrective actions.

The key issue in assessing the success of certification is not the absence of failures. This is an unrealistic and impractical measure of certification's performance. A more useful question to ask is, are there significantly fewer failures exhibited by certified operations when compared with conventionally managed operations? Secondly, what are the severity and impact of these failures, and are they generally less severe (with a reduced or more limited impact) than those that occur outside of certified areas? Finally, what represents an 'acceptable' level of failures within certified organisations?

Unfortunately, there are no hard data readily available to permit this sort of objective analysis. Data is maintained by certification schemes and their CBs and by the accreditation bodies that provide the technical and management oversight of the schemes. The number of inactive and withdrawn certificates is available. However, failures of certificate holders do not always result in cancellation or withdrawal of certificates. Producing a detailed breakdown of certification 'failures' would require an in-depth analysis of each certificate holder's historical performance. The reports of CBs include a record of the corrective actions issued by them for each certificate holder. Accreditation body reports on CBs also maintain similar summaries of the types of corrective actions issued and this could provide the raw material for such an analysis, but this information is typically confidential.

Many disputes, complaints and challenges, including system failures, are resolved through a series of internal processes between certificate holder, CB and certification scheme before they become public knowledge. The actions of the watchdog and monitoring organisations, and the evidence that they have accumulated and that they release directly into the public domain, usually activates the relevant dispute resolution process. The final decisions of this process may still be challenged by those who have initiated the complaint, but the key point is that the complaint has been subjected to due process. If the activities can be shown to be illegal or resulting in significant harm, then a legal challenge can be mounted. However, even watchdog and monitoring organisations with their technical skills and financial resources have found it very difficult to achieve a conviction in court.

There have been a number of attempts to compare the relative strengths of different schemes such as the FSC and the PEFC. Probably the comparison between certification schemes that has exerted the greatest influence has been the successive comparisons conducted by the UK government's Central Point of Expertise on Timber (CPET). CPET was established in 2000 to provide independent guidance to the UK government on the procurement of sustainably produced timber, and this has guided the purchasing strategies of many of UK's commercial processors and importers. CPET utilised the UK government's own criteria for assessing the legality and sustainability of timber, though these did not include criteria to assess social or cultural values. Such limitations in the criteria used have damaged CPET's credibility among some NGOs and others.

In 2004 and 2009 CPET conducted an assessment of a number of competing schemes, including the FSC and the PEFC. The assessment found the FSC and the PEFC schemes offered similar levels of assurance that the timber certified by the schemes met the UK's criteria for legality and sustainability of timber. It rated two other schemes (CSA and SFI) as meeting the criteria but with a lower level of assurance. These schemes have subsequently been endorsed and subsumed within the PEFC. After the 2004 assessment many commercial companies expressed their concerns with the findings and continued to believe that the FSC scheme represented a superior certification with greater credibility (Benfield, 2005).

In its 2009 assessment, CPET did suggest the FSC scheme was marginally superior, scoring 47 for FSC and 41 for the PEFC. However, the absence of any social or cultural criteria effectively eliminated the areas, which many considered were the FSC's principal strengths and the reason that the scheme was preferred over all others. The FSC included specific criteria and indicators for key social and cultural parameters, including C & I related to the rights of indigenous and native peoples, which was felt to represent a more comprehensive definition of 'sustainability'.

By this point it was clear that the broad equivalence of the two global schemes in terms of commercial acceptability had become an accepted fact of life. This effectively eliminated the FSC's principal commercial advantage of access to markets demanding 'sustainability' in sourcing. This contributed to an accelerated uptake of PEFC-endorsed certifications (Fig. 1). The PEFC overtook the FSC and became the scheme with the largest certified area in 2006. The 2015 CPET findings have reinforced the primacy of the two global schemes – which are now classified as meeting 100% of legality criteria and 90% of the social criteria (CPET, 2015; PEFC, 2015).

Nevertheless, certain NGOs have cast doubt on the credibility of these findings. The CPET analysis of FSC certification can be compared, for example, with those of NGOs such as Nature People Economy Connected (NEPCon), which has identified weaknesses in the way the FSC ensures traceability in the supply chain, including tracking volumes of timber traded, mixing of certified and non-certified timber and effectively identifying illegal sources of timber (NEPCon, 2019). NEPCon has also highlighted the problems faced by community-based forest management groups in navigating the complexities and costs of the FSC scheme (NEPCon, 2018).

An alternative approach to comparing certification schemes would be to adopt a subset of criteria and indicators that the schemes have in common, which with a little modification could provide the basis of a scoring chart against which all schemes could be compared. This is broadly the approach that has been adopted for a 2019 analysis of the contribution of REDD+ and FLEGT-VPA processes to the achievement of SFM in Indonesia (Neupane et al., 2019). A meta-analysis of 26 studies by Clark and Kozar (2011) comparing schemes such as FSC and SFI found that FSC scored best in terms of ecological and social sustainability indicators. The problem is in finding indicators that are genuinely objective. The Clark and Kozer study based its comparisons on discussions and structured interviews that were inevitably more subjective.

The broad conclusions of this study were that certification has contributed to SFM through ensuring increased transparency, the application of standards that ensure the more active participation of a wider range of stakeholders, particularly those of indigenous peoples and forest-dwelling (or forest-dependent) communities, and the 'institutionalisation' of enhanced techniques for harvesting and forest protection. But the study also found that stakeholders have expressed a general dissatisfaction with a failure to address key issues, a number of which could be regarded as key elements of sustainability. These include resolving long-term rights related to ownership and tenure and the failure to adequately empower disenfranchised stakeholder groups. Other research has highlighted the contribution of FSC and other certification schemes to promoting SFM practices (Pena-Claros et al., 2009).

7 Conclusion

European and North American companies have been best placed to exploit the commercial advantages of forest certification because their markets were more open to its rationale. Initially, certification offered preferential access to specific markets and companies reconfigured their supply chains to enable them to meet the demands of certification. As certification became an increasingly desirable and sought-after commodity, importers, processors and manufacturers were able to prevail on their suppliers to meet the demands of certification. Mechanisms include exclusivity contracts, shared costs of certification, business to business contracts and other contractual arrangements. Suppliers that could more easily meet the requirements represented 'the low hanging fruit' from which the FSC, as the first scheme to be commercially available, was able to benefit.

The PEFC was designed initially as a European scheme – although this changed in 2003 – and its certification was aimed initially at satisfying its Scandinavian supporters. Similarly, the SFI and CSA schemes were designed to benefit North American suppliers and permit them to enter the certification market place. The early adoption by European and North American suppliers and the failure of tropical suppliers to take advantage of the opportunities provided by certification have thus contributed to this uneven distribution of certified forests.

Many observers and stakeholders have expressed disquiet about the domination of certified forest areas by North American and European forests, and the small proportion of forest areas that had been certified in Africa, Asia, South America and Africa, and from the tropics in general. Areas of certified forest in Asia have hovered around 3-5% of the total certified area for the past 10 years, in spite of Asia containing three of the biggest timber producers by volume in the world (Indonesia, Malaysia and Papua New Guinea) and three of the most sensitive sources (Myanmar, Laos and Cambodia). Asia also includes some of the world's largest markets (China and India) through which processed timber products are then supplied to Europe and North America.

The available data reveals that forest certification has successfully established itself as an integral part of forest management and global supply chain management, particularly in temperate and boreal forest areas, but much less so in tropical forests. Forest certification has also become a viable commercial business in its own right and has contributed to the changes that the international timber trade has undergone over the last 25 years. However, the impact of forest certification in terms of bringing about improvement in the management of the world's forest resources is much more difficult to establish objectively.

The data and metrics most widely available - numbers of certificate holders, area of forest, volume of timber as a proportion of total trade and so on - are not adequate surrogates by which to assess the level of sustainability achieved or the uptake of 'good forest practice.' They are insufficiently precise and are subject to considerable interpretation. Simpler and more objective performance indicators are required to monitor changes in key forest parameters at a certificate holder level in order to establish with confidence the levels of sustainability achieved by individual certificate holders.

The beneficial impacts of forest certification will not be distributed evenly between the certification schemes, so presenting an aggregate of the benefits can lead to a misunderstanding of how and why these benefits have been achieved. Establishing the true impact should be based on an assessment of the performance of certificate holders over time, scheme-by-scheme. This assessment should be conducted against a set of objective performance indicators reflecting key forest parameters: only then will it be possible to establish objectively how effective specific certification schemes have been in contributing to the pursuit of the twin objectives of establishing compliance and promoting improvements and change towards SFM. Data is more generally available on the first objective. However, forest certification, through the benefits it provides principally market access and the networking of suppliers and customers – is intended to drive meaningful changes in forest practise: this has yet to be demonstrated conclusively.

The influence of ISO and related international systems and processes has resulted in forest certification systems becoming increasingly similar in operational practise. ISO systems now determine many key aspects of a forest certification scheme's operational practice including, accreditation, audit practice and process, auditor competence and training and reporting. The standardisation of approach has clearly helped ensure forest certification acceptance by the commercial forestry and timber sectors, but it is also clear that this has limited the effectiveness of the individual schemes to deliver their unique visions of SFM.

A fear expressed by a number of stakeholders is that the drive for conformity (with all schemes adopting a common aim and delivery mechanism) will inevitably result in the adoption of a form of SFM that represents the lowest common denominator rather than an ideal. This sort of impact can be seen clearly within the FSC, the 'original' forest certification scheme, and this has resulted in a number of high-profile NGOs and supporters withdrawing their support from it, citing a loss of confidence with the FSC's ability to deliver on its promise of ensuring genuine, credible and rigorous, SFM.

After 24 years since the FSC was started, this remains the case and the continuing levels of deforestation make clear that overcoming deforestation and non-sustainable practices will require certification to be reinforced by national and international laws that have real teeth and that these regulations have to be applied vigorously, with commitment and sufficient resources. The Lacey Act in the United States and the EU's Timber Regulation of 2013 (EUTR) show how effective legal and regulatory measures can be, if structured properly.

The other missing ingredient is commitment. Regulators, politicians, managers and government officers all have to be committed to a common vision. Without this, no enterprise will be able to consistently achieve its goals. Forest certification seeks to eliminate the need for an appropriate moral commitment, and replaces it with the achievement of clear and tangible physical goals, which are then rewarded by certain commercial benefits. As explained in the analysis, it is clear that this may not provide sufficient incentive to ensure consistently an organisation's compliance with the standards.

Finding evidence to demonstrate beneficial impact on rates of deforestation and forest degradation from certification is difficult, particularly in tropical forests where forest certification's uptake is clearly well below that in temperate forests and boreal forests. Instead, what is evident are specific improvements in the management of forests within certified areas. These include: improved transparency; enhanced levels of participation in on-the-ground management; more equitable income distribution within local communities; the creation of wildlife corridors, buffer zones and areas of protection and conservation within forest concessions. These improvements are frequently cited in the annual reports of the principal certification schemes.

Certification's 'success' is not – or should not – be determined by its utility to the commercial sector. Certification was originally designed as a tool to identify and reward those organizations which were practising SFM, and although it will inevitably continue to develop and change in response to external pressures – commercial, political, environmental and social – its relevance and acceptance is dependent on it delivering on its primary goal, and to do this it has to maintain the trust and confidence of all its stakeholders in respect of the assurances that certification provides.

This stakeholder base clearly includes those directly responsible for growing, processing and manufacturing timber products, but it also includes those purchasing and using those products – the customers or end users. Within many supply chains this key group of stakeholders have become increasingly demanding, but in the timber industry their expectations remain relatively straightforward and consistent: the expectation is that certification delivers sustainably managed forest resources.

Certification's contribution as an active 'driver' of SFM has clearly diminished with the expansion of the schemes and diversity of certification standards, but it is critical that certification performs its prime function diligently and consistently, by providing the objective assurances required by those stakeholders committed to SFM and through which they are able to exercise their purchasing preferences. It is vital, therefore, that certification retains the trust and confidence of all its stakeholders; without it the value of certification is significantly diminished and the benefits of sustainable forest management are more difficult to discern and promote.

8 Where to look for further information

A wealth of data and information is available from a wide range of sources. The forest certification schemes generate much of this themselves, and the bulk of this is widely available, either on-line through their dedicated websites or published as hard copy in the form of reports and technical publications. These sources provide scheme-specific data but information about the sector as a whole – collated and presented for all forest certification schemes – is less common. The PEFC and the FSC, as the two largest global schemes, frequently provide data and information in order to compare the progress of both schemes, and data is presented regularly on the numbers of certificate holders, geographic coverage and the area of forest covered by certificates.

As the principal North American schemes have been endorsed by the PEFC, their data has been incorporated within the PEFC's databases so the

PEFC website provides a very simple means of comparison. Data relating to the FSC has been extracted from FSC websites, principally: https://fsc.org/en/page/facts-figures and its interactive map: https://fsc-int.maps.arcgis.com/apps/webappviewer/index.html?id=06188ad39e5344db96a4a181e135c393. Data relating to the PEFC has, likewise, been extracted from its websites: https://www.pefc.org/discover-pefc/facts-and-figures.

Data is also produced by international agencies including the FAO and the European Union. The FAO's 5-yearly Forest Resource Assessments (http://www.fao.org/forest-resources-assessment/documents/en/) and the Annual State of the World's Forests (http://www.fao.org/state-of-forests/en/) are useful sources for macro-data, but the data in these documents frequently does not coincide with data generated by other organisations. Discrepancies are particularly evident in the case of data related to forest areas, forest loss and degradation. Additionally, data is often presented in formats that require further data collection and analysis before they can be fully understood: for example, forest areas are presented as a percentage of land area, but forest loss is presented with no reference to the original forest area. This can be confusing and does not permit easy comparison or analysis.

More useful are the data on global and national deforestation and degradation captured, collated and re-presented by organisations, which disseminate forestry and timber-related information. Good examples are Mongabay (https://rainforests.mongabay.com/deforestation/archive/) and Global Forest Watch, which is part of the World Resources Institute (https://www.globalforestwatch.org/howto/view-statistics/statistics-gfw-view-country-rankings.html). Both organisations publish information specifically related to forest degradation and destruction and have been a source for this chapter, but it is important to understand that this information is sometimes presented selectively or is already formatted to support a particular conclusion. Additionally, the data does not coincide with that presented through the FAO publications. Nevertheless, these are an excellent source of pre-compiled data.

The CBs also provide considerable information and statistics on their forest certificates and certificate holders, and are obliged to do so by the terms of their accreditation contracts. As many CBs conduct certification audits on behalf of different schemes, scheme-specific data can sometimes be a little difficult to extract. Other material – especially commercially sensitive information relating to pricing and sourcing and the relationship between scheme owner and its CBs and the CBs and the certificate holders – are subject to greater levels of security and are generally not available to the general public or only on special request. This applies to information related to disputes and challenges relating to certificates and their cancellation and withdrawal. This information is usually presented by the scheme owner.

9 References

Arets, E. J. and Veeneklaas, F. R. 2014. Costs and benefits of a more sustainable production of tropical timber. WOt-technical report 10. Wageningen University, The Netherlands.

Auld, G. 2014. *Constructing Private Governance: The Rise and Evolution of Forest, Coffee, and Fisheries Certification*. Yale University Press, New Haven.

Bass, S. 1996. Introduction to forest certification. Discussion Paper 1. European Forest Institute (EFI), Joensuu, Finland. Available at: https://www.efi.int/projects/certification-information-service-cis.

Bass, S. 1998. Forest certification – The debate about standards. Rural Development Forestry Network (RDFN), Paper 23b. Overseas Development Institute, London, UK.

Benfield 2005. The truth About FSC and PEFC certification. Benfield ATT Group. Available at: http://www.benfieldattgroup.co.uk/knowledge-centre/advisory/environment/timber-certification/.

Blaser, J., Sarre, A., Poore, D. and Johnson, S. 2011. Status of tropical forest management 2011 [Online]. International Tropical Timber Organisation. Available at: https://www.itto.int/direct/topics/topics_pdf_download/topics_id=2660&no=0&disp=inline (accessed 20 June 2019).

Brack, D. 2013. Combatting illegal logging: Interaction with WTO rules. Chatham House Briefing Paper EER BP 2013/01. Available at: https://www.chathamhouse.org/sites/default/files/public/Research/Energy,%20Environment%20and%20Development/0513bp_brack.pdf.

Burivalova, Z., Hua, F., Koh, L. P., Garcia, C. and Putz, F. 2017. A critical comparison of conventional, certified and community management of tropical forests for timber in terms of environmental, economic and social variables. *Conservation Letters* 10(1), 4–14. doi:10.1111/conl.12244.

Burley, F. W. 1988. The tropical Forestry Action: Plan recent progress and new initiatives. In: Wilson, E. O. and Peter, F. W. (Eds), *Biodiversity*, National Academic Press, Washington, DC, USA. https://www.ncbi.nlm.nih.gov/books/NBK219314/.

Cashore, B., Auld, G. and Newson, D. 2004. *Governing through Markets: Forest Certification and the Emergence of Non-State Authority*. Yale University Press, New Haven, CT.

Cashore, B., Gale, F., Meidinger, E. and Newson, D. (Eds) 2006. *Confronting Sustainability: Forest Certification in Developing and Transitioning Countries*. Forestry and Environmental Studies Publication Series (No. 8). Yale School of Forestry, New Haven, CT.

Changing Markets Foundation 2018. The false promise of certification. Available at: https://www.changingmarkets.org/wp-content/uploads/2018/05/False-promise_full-report-ENG.pdf.

Clark, M. R. and Kozar, J. S. 2011. Comparing sustainable forest management certification standards: a meta-analysis. *Ecology and Society* 16(1), 3–10. doi:10.5751/ES-03736-160103.

CPET 2015. Timber procurement policy (TPP). Central Point of expertise on timber (CEPT). Department for Business – UK Government. Available at: https://www.gov.uk/government/groups/central-point-of-expertise-on-timber.

Doyle, T. and MacGregor, S. (Eds) 2013. *Environmental Movements around the World: Shades of Green in Politics and Culture*. Praeger Publishers, Santa Barbara, CA.

Elliott, C. 2000. *Forest Certification: A Policy Perspective*. CENTER for International Forestry Research, Bogor, Indonesia.

Gatti, R. C., et al. 2019. Sustainable palm oil may not be so sustainable. *Science of the Total Environment* 652, 85–51.

Greenpeace 2018. Greenpeace International to not renew FSC membership. Available at: https://www.greenpeace.org/international/press-release/15589/greenpeace-international-to-not-renew-fsc-membership/.

Humphreys, D. 1996. *Forest Politics: the Evolution of International Cooperation*. Earthscan Publications, Abington, Oxon, UK.

Humphreys, D. 2006. *Logjam: Deforestation and the Crisis of Global Governance*. Earthscan Publications, Abington, Oxon, UK.

Jakub, M. 2019. Analysis of socio-economic impacts of FSC and PEFC certification on business entities and Consumers. *Sustainability* 11, e4122.

Klabbers, J. 1999. Forest certification and the WTO. Discussion Paper 7. European Forest Institute (EFI), Joensuu, Finland. Available at: https://www.efi.int/projects/certification-information-service-cis.

Kleinschmidt, D., Mansourian, S., Wildburger, C. and Purret, A. 2016. Illegal logging and related timber trade - dimensions, drivers, impacts and responses. IUFRO World Series (Vol. 35). Available at: https://www.iufro.org/science/gfep/illegal-timber-trade-rapid-response/report/.

Lammerts von Bueren, E. and Blom, E. 1997. *Hierarchical Framework for Formulation of SFM Standards: Principles, Criteria and Indicators*. The Tropenbos Foundation, Wageningen, The Netherlands.

McDermott, C. 2011. Trust, legality and power in forest certification. *Geoforum* 43(3), 634–44.

Muhtaman, D. R. and Prasetyo, A. P. 2006. Forest certification in Indonesia. In: Cashmore, B., et al. (Eds), *Confronting Sustainability: Forest Certification in Developing and Transitioning Countries*. Forestry and Environmental Studies Publication Series (vol. 8). Yale School of Forestry, New Haven, CT.

NEPCon 2018. Alternatives to Facilitate FSC Certification for Community Forestry Operations. Nature Economy and People Connected (NEPCon). Available at: https://www.nepcon.org/library/report/alternatives-facilitate-fsc-certification-community-forestry-operations-cfe.

NEPCon 2019. How Forest Schemes Meet the European Timber Regulation (EUTR) Requirements: Forest Stewardship Council (FSC). Available at: https://www.nepcon.org/library/report/certification-schemes-report-fsc.

Neumayer, E. and Perkins, R. 2004. What explains the uneven take-up of ISO 14001 at the global level?: a panel-data analysis. *Environment and planning A* 36(5), 823–39. ISSN 0308-518X, doi:10.1068/a36144.

Neupane, P. R., Wiati, C. B., Angi, E. M., Köhl, M., Butarbutar, T., Reonaldus, P. and Gauli, A. 2019. How REDD+ and FLEGT-VPA processes are contributing towards SFM in Indonesia – the specialists' viewpoint. *International Forestry Review* 21(4), 460–85. doi:10.1505/146554819827906807.

Newsom, D. and Hewitt, D. 2005. *The Global Impact of SmartWood Certification*. Rainforest Alliance, Richmond, VT.

Pearce, D., Putz, F. and Vanclay, J. 1993. *A Sustainable Forest Future?* Centre for Social and Economic Research on Global Environment: CSERGE Working, pp. 99–15, Paper GEC.

PEFC 2015. CPET update gives forest certification schemes top scores. Available at: https ://www.pefc.co.uk/news_articles/cpet-update-gives-forest-certification-schemes-t op-score.

PEFC Annual Report 2019. Available at: https://www.pefc.co.uk/system/resources/ W1siZiIsIjIwMTkvMDUvMjAvM3I2ejFyYjJpcl9QRUZDX0FOTlVBTF9SRVBPUlRfMj AxOV9XRUIucGRmIl1d/PEFC%20ANNUAL%20REPORT%202019%20WEB.pdf.

Pena-Claros, M., Blommerde, S. and Bongers, F. 2009. *Forest Management Certification in the Tropics: an Evaluation of Its Ecological, Economic and Social Impact*. Forest Ecology and Forest Management Group, Wageningen University, The Netherlands.

Petera, P. and Vlosky, R. 2006. History of forest certification. *Louisiana Forest Product Development Center*. Working Paper 71. Louisiana State University, USA.

Rametsteiner, E., et al. 1999. Potential markets for certified forest products in Europe. Discussion Paper 2. European Forest Institute (EFI), Joensuu, Finland. Available at: https://www.efi.int/projects/certification-information-service-cis.

Thornber, K., Plouvier, D. and Bass, S. 1999. Certification - Barriers to benefits: Discussion of equity implications. Discussion Paper 8, European Forest Institute (EFI), Joensuu, Finland. Available at: https://www.efi.int/projects/certification-information-service-cis.

Wang, S. 2001. Towards an international convention on forests: building blocks and stumbling blocks. *The International Forestry Review* 3(4), 251–64.

Winterbottom, R. 1995. The tropical forestry action plan: is it working? *NAPA Bulletin* 15(1), 60–70. doi:10.1525/napa.1995.15.1.60.

WWF 2014. Research review: the impact of Forest Stewardship Council (FSC) certification. Available at: https://wwf.panda.org/?231170/Research-review-The-impact-of-Fores t-Stewardship-Council-FSC-certification.

Chapter 4

Sustainable forest management (SFM) of tropical moist forests: the Congo Basin

Paolo Omar Cerutti and Robert Nasi, Center for International Forestry Research (CIFOR), Kenya and Indonesia

1   Introduction

2   Logging concessions, the land and management plans

3   Land zoning

4   Sustainable forest management plans

5   From 'nesting dolls' to improved policies

6   Improving processes and institutions

7   From theory to practice

8   Conclusions

9   References

1 Introduction

The Congo basin contains the second largest expanse of tropical forests on the planet, after the Amazon, and it includes six countries, namely Cameroon, Central African Republic (CAR), Equatorial Guinea, Gabon, Republic of Congo and Democratic Republic of Congo (DRC). In 2018, the six countries had a population of about 122 million, estimated to more than double by 2050, at about 274 million (UN-DESA, 2018). The same countries cover an area of about 400 million ha, of which about 187 million ha are dense moist forests and about 82 million ha are other types of forests, such as mosaics, dry forests and savannah (de Wasseige et al., 2013). The DRC stands out among other countries for both its population and forested area: about 84 million people (more than double the population of all the other countries combined), soon to become 197 million, and about 62% of total forested areas, or 166 million ha (Table 1).

Sustainable Forest Management (SFM) is looked at in this document through the lenses of the past (e.g. its origins in the region and its current situation in terms of implementation) but with a keen interest in the future. The

http://dx.doi.org/10.19103/AS.2020.0074.41

Table 1 Population and forest area

Country	Population 2018 ('000)	Population 2050 ('000)	Sup. dense forests (ha)	Sup. other forests (ha)
Cameroon	24 678	49 817	19 091 044	7 763 961
Central Africa Republic	4737	8851	6 923 690	15 324 776
Democratic Republic of Congo	84 005	197 404	114 526 051	52 134 555
Equatorial Guinea	1314	2845	2 120 060	507 453
Gabon	2068	3516	22 505 397	1 487 747
Republic of Congo	5400	11 510	21 278 180	4 479 501
Total	122 202	273 943	186 444 422	81 697 993

near future in particular and what political decisions it will carry are key to the fate of the Congo basin's forests.

West African countries – where SFM has been the objective of forest policies for several decades – are often used as (dire) examples of what could possibly happen to the Congo basin in the coming years, as forests get depleted in the former while logging companies move South-East towards the latter. Cote d'Ivoire's forest cover went from 16 million ha in the 1960s to less than 3 million ha in the late 1990s (Carrière, 2011). Today, when one looks at a map of the country indicating forest use, one can still see about 384 'forest exploitation perimeters' covering the entire southern part of the country, officially sustainably managed through a management plan, when in fact most of those areas have been agricultural fields for years with scattered pockets of remaining trees. In Ghana, the current rate of timber extraction could be as high as six times the (sustainable) annual allowable cut (Hansen et al., 2012).

No doubt, direct and indirect causes for trends of deforestation may have been different in West Africa. For example, active political planning and large public and private investments on agricultural expansion to support cocoa production, with direct impacts on infrastructure, come to mind in the case of Cote d'Ivoire and Ghana. These factors are still largely lacking in the Congo basin. Yet things may rapidly change. So far, large-scale investments in industrial agriculture have been scant when compared to those occurring in other remaining tropical regions such as the Amazon or Southeast Asia, but all countries in the basin plan to sustain their paths to development through increased agricultural production and large-scale investments (Conigliani et al., 2018; Ordway et al., 2017a).

Several other factors need consideration because of their past and potential future direct and indirect impacts on forests. First, the tenurial system in all countries of the Congo basin *de jure* leaves only nihil to tiny percentages of the forest in indigenous or communal hands. In countries such as Cameroon, Gabon and, more recently, the DRC there have been attempts to at least devolve some forest management rights to local communities - notably through the institutionalisation of 'community forests' - but the State maintains the upper hand on the vast majority of the remaining forest (Rights and Resources Initiative, 2015), which may be conducive to people clearing more forest to assert *de facto* ownership (Ordway et al., 2017b).

Second, past and present conflicts and wars, with millions of internally and externally displaced people in need of - among many other urgent things - wood for their energy and cooking needs, have not only contributed to some of the lowest scores on human development but they have also often taken a toll on forested areas officially aimed at the implementation of SFM (e.g. refugees from CAR into Eastern Cameroon, or from Nigeria into Western Cameroon or in several areas of the DRC), with logging companies often being directly questioned about their roles and responsibilities in such situations (e.g. Debroux et al., 2007; Global Witness, 2015; or https://www.greenpeace.org/africa/en/press/1929/greenpeace-africa-expresses-shock-over-wijma-activit ies-in-cameroon/).

Lastly, all of the above in addition to political conditions often verging towards State failure (Herbst, 2000), notably in areas away from major cities where forests are located, have contributed to - among others - road, rail and electricity networks remaining poorly developed or non-existent, resulting in very high cost of transport and production for both public and private efforts at SFM implementation.

With all this in mind, the question should be asked today as to whether the 'forest sector' in the countries of the Congo basin - in particular where SFM is to be implemented and forests are supposed to remain forests - will be just a passive recipient of more powerful vectors of change, or whether it could withstand them, and indeed help steer them through more coherent, viable and sustainable business propositions. The remainder of this chapter attempts to provide some elements of response to that question, by discussing existing trends on SFM and some ways forward.

First, some information on logging concessions will be presented. Historically, there has been a tendency to identify the concept of SFM with the practical instrument through which SFM is supposed to be implemented on the ground inside logging concessions, that is with the forest management plan (Nasi et al., 2006). Over the years, for example, several international organisations have pointed to the efforts made by governments in the Congo basin towards improved sustainability, explicitly suggesting that a growing number of approved forest management plans should be read as an indicator of improved sustainable management (CBFP, 2006; COMIFAC, 2004). The next section will thus present the current situation with logging concessions that should be managed according to a management plan. Though starting with a quantitative presentation, the section will however suggest that a broader understanding of SFM is needed if better impacts on the ground are to be expected.

Such broader understanding will then be discussed in more detail in the following section, which will turn to a more qualitative explanation of past and current trends in the implementation of SFM. Overlapping the layers of local biophysical, human and political conditions, there have been international processes, financial shocks, and their regional and national institutional responses. These must also be considered in order to better frame and understand the trends in the adoption and implementation of SFM over the past four decades in the Congo basin, as well as to better gauge - and hopefully coherently and actively plan - its future. In this section, some of the most relevant processes and institutions will be presented.

The last section will summarise the main findings, make a few suggestions for the future of SFM in the Congo basin and conclude.

2 Logging concessions, the land and management plans

The concept of SFM in the Congo basin, as elsewhere, is very much related to land classification and tenure systems and to the planned objectives of different land categories. The literature abounds on detailed analyses of all categories in which the forests of the basin can be classified (for a comprehensive legal description, see CBFP, 2007, Chapter 8). Suffice here to say that all countries allocate logging permits mandated to implement SFM into a particular category of (publicly owned) forests which can be broadly grouped under the term 'production forests'. In turn, though the specific terminology varies from country to country, production forests generally fall within variously defined 'permanent forest domains', similar to national parks and some other types of protected areas.

The most common logging permit granted to access production forests are logging concessions, whereby a private entity is given permission to manage a public property. While the term concession broadly refers to the legal agreement itself, the names of the actual forest areas granted to private companies vary. In Cameroon, for example, it is common to speak of forest management units (FMU) and each company cannot be granted FMUs exceeding a total area of 200,000 ha, while in the CAR the term would be exploitation and management permits, granted to private companies with no sealing on forest area, as in neighbouring Republic of Congo. All these areas are mandated by law to be managed through a forest management plan.

The most recent numbers available for the six countries of the Congo basin indicate that about 370 concessions exist, covering a total area just short of 50 million ha (Table 2). On average, about 24 million ha or about half (47%) of those concessions are today harvested with officially approved forest management plans. Cameroon and CAR stand out for their high percentage of concessions with approved plans (88% and 82%, respectively), followed by Gabon (67%) and Republic of Congo (40%). The DRC just recently started approving management plans, lagging very far behind other countries (7%).

Insofar as SFM is concerned, Cameroon deserves a particular mention because in addition to logging concessions, it allows so-called council forests to be treated as *de facto* concessions, and mandates them to prepare and obtain management plans. As of April 2019, Cameroon had about 36 approved council forests covering an area of about 840 000 ha. A similar situation, in terms of potential areas covered, though with different terminology and under very different legal conditions, may soon become a reality in DRC, whereby forests of the communities up to a maximum of 50 000 ha each, may be managed as *de facto* concessions (Vermeulen and Karsenty, 2017).

In summary, and ideally as the 'pyramidal' logic of the legislator goes, permanent domains are created which contain production forests, which in turn contain logging concessions which must be managed following a forest management plan approved by the relevant ministry. If the logic holds and going back up the pyramid, management plans would thus guarantee logging concessions (and council forests in the case of Cameroon) to be managed sustainably and production forests to remain both productive and forests, thus fulfilling their mandate as permanent forest domains of the State.

Both the logic and the terminology are extremely important because they are based on an implicitly top-down system of forest governance, whereby land-use planning is conducted at the national level, with a central government capable and willing to convene – and ideally deliberate with – all relevant parties (e.g. private sector and civil society), sectors (e.g. agriculture, forests, mining, territorial administration), ministries and decentralised government units. In turn, once a national master plan on permanent forests and their sustainable use is agreed upon, various ministries and decentralised units will implement it within the boundaries of their relevant geographies, powers and responsibilities. This also means that, for example, the Ministry of Agriculture will know where those permanent forests are and will not grant land to, say, agribusinesses on those same lands, or at least not until a full consultation and agreement have been conducted with other relevant ministries and/or the industrial operator to whom those forests have already been granted.

In fact, such logic has never had much to do with the reality of forest governance in the Congo basin. Two examples illustrate this reality. First, no

Table 2 Forest concessions (various years)

Country[a]	Concessions			Managed concessions		
	Area (ha)	Number	Average area	Area (ha)	%	Certified (FM)[b]
Cameroon	6 281 212	105	59 821	5 522 682	88	341 708
Congo	13 913 699	50	278 274	5 555 629	40	2 410 693
Gabon	14 197 038	97	146 361	9 469 504	67	2 042 616
Equatorial Guinea	740 122	48	15 419	0	0	
CAR	3 698 531	14	264 181	3 023 880	82	
DRC	10 762 055	57	188 808	775 713	7	
Total	49 592 657	371	158 811	24 347 408	47	4 795 017

[a] All information from https://www.observatoire-comifac.net/monitoring_system/concessions.
[b] At the time of writing, only FSC certificates had been delivered in the Congo basin.

country has ever completed a comprehensive land-use zoning plan. Second, because of the weak institutional landscape, all countries have more or less deferred their responsibilities to prepare forest management plans to the private logging companies to which logging concessions were granted. These remain major obstacles to the effective implementation of SFM, and they deserve more detailed explanations, presented below.

3 Land zoning

The issue that needs mention here because of its potential impacts on the implementation of SFM – and because it often leads to confusion when discussing forests and tenure in the Congo basin – is the division between the proper zoning that may be conducted to attribute particular uses to the land (e.g. forest, agriculture, mining), and the attribution of rights to physical or moral persons (e.g. the State, communities, private individuals or companies). Within this distinction, only Cameroon can claim to have undergone a real zoning at the beginning of the 1990s, with the involvement of various ministries and interest groups (Côté, 1993), though the process is still incomplete about thirty years later, so far restricted to the Southern part of the country, and very much focused on delimiting the boundaries of logging concessions over a total area of about 6 million ha (MINEF, 1995).

Several other countries have been conducting similar attempts, but without much success on either partial or final implementation to date. Eventually, all countries ended up with logging concessions demarcated in official maps, harvested by logging companies, and paying much needed revenues to States which, in return, could only promise that official zoning would occur sometime in the future. In other words, in some countries the designation of logging concessions and their attribution to private entities has been considered as *de facto* zoning (Karsenty, 2018), while waiting for their *de jure* recognition.

This remains a most contentious issue because if there is no land-use master plan, the attribution of rights is weakened as different rights can be attributed by various ministries with varied interests. It is indeed not reassuring to know that already about 30% of the total surface of current logging concessions overlaps with mining exploration permits, with values as high as 54% in Gabon (de Wasseige et al., 2013). Exploration does not always lead to discoveries and exploitation contracts. But this is immaterial insofar as the philosophy and long-term thinking behind SFM are concerned. What matters in such a situation is that the perspective of an investor changes radically: on the one hand, a site manager is asked to invest in the preparation of management plans which should be of high quality (hence very expensive) if useful for serious management purposes; on the other hand, the same site manager is concurrently sent the message that those investments may be for nothing after all, because a variable percentage of the granted concessions may become the site for an oil, gold, oil palm, rubber or other investment.

This has been a continuous and growing issue especially in the last couple of decades, with land-use conflicts arising in many countries on the forest-agriculture-mining interface (e.g. SFIL, 2015; Greenpeace International and The Oakland Institute, 2013). Such issues will become even more contentious as countries adopt and will hopefully implement legislation aimed at granting communities management and land rights over their ancestral lands.

4 Sustainable forest management plans

The forests of the Congo basin were exposed to notions of sustained yields and multiple-use forestry since the first half of the twentieth century, notably with experimental forestry occurring in famous research stations such as the Yangambi forest reserve in DRC. Those notions morphed over time to become what we know today as SFM (Nasi and Frost, 2009), and eventually entered the arena of forest policies and regulations in the Congo basin through the discussions that took place around the revision of the 1981 Forest Law of Cameroon. The revision took place in 1988 under the auspices of the Tropical Forest Action Plan (TFAP). Various topics – including a transparent auction for logging concessions, fiscal reforms, and mandatory forest management plans – were vigorously supported by the World Bank through structural adjustment plans' (SAP) conditional lending, and ultimately made their way into what became the 1994 Forest Law and its 1995 implementing decree, notwithstanding a strong resistance from various parties (Brunner and Ekoko, 2000; Ekoko, 2000; Gros, 1995).

Of particular interest are the interpretations and modifications, again *de facto* if not *de jure*, of the articles concerning the responsibilities around the adoption and implementation of SFM. Article 64 of the Cameroon forest law (Republic of Cameroon, 1994) states: 'Forest management shall be the concern of the ministry in charge of forests working through a public body. It may subcontract certain management activities to private or community bodies.' In the 1995 Decree, the responsibilities to (i) conduct the management inventory and (ii) prepare the management plan are more clearly removed from the 'public body' and given to the same concessionaires to which logging concession would later be attributed (Republic of Cameroon, 1995).

The delegation of responsibilities – on par with the situation of land zoning described above – is very important for SFM because it was introduced for the first time in the Congo basin and, since then, it has influenced in a way or another all other forest legal frameworks or their actual implementation in the entire region (Karsenty, 2006). One notable exception remains the CAR, whereby the limited number of logging concessions (14) and a management unit funded by the French Development Cooperation have allowed management plans to be prepared by the relevant ministry, albeit still with the active participation and supervision of logging companies.

In practice, logging concessions are attributed to private companies who promise to prepare forest management plans guaranteeing the sustainable use of the resource. In turn, forest officials should review, verify, approve and regularly monitor the implementation of the plans.

There are several problems with this system, starting with the distinction between the duration of the contractual agreement linking the State and the private entity (i.e. the concessionaire) and the duration of the management plan. While in CAR the concession is granted for the entire lifespan of the company, in all other countries there exist legal temporal limitations to the contractual agreement, from a minimum of 15 years in Cameroon, Republic of Congo and Equatorial Guinea, to 25 years in DRC and 30 years in Gabon.

While in theory concessions are renewable, in all countries bar CAR companies are thus not legally guaranteed that their initial investments (e.g. management plan, sawmills, engagements with local populations) can be spread beyond the initial contract. As is the case with insecure zoning rights, this is very important in setting the mindset and investment horizons of logging companies, both remaining mostly aimed towards the short-term. Management plans mandated by law aim, among others, at establishing a *sustainable* rotation period, that is the number of years between one cut and the next in the same forest area. In turn, this will guarantee that the functions and services offered to humanity by the forests of today will be similar to those offered by the (harvested) forests of tomorrow.

Yet in such a system one can naturally expect that the length of the contract will trump any consideration about the sustainability of the management plan. The result is that logging companies will tend to prepare management plans that broadly overlap with the contractual agreement. This is why most of the existing management plans in the Congo basin generally set the rotation at about 25-30 years. Given a good knowledge of the forest, the correct assumptions and the right techniques, rotation can of course be set at 25–30 years. Under the current conditions, however, we argue this is done less to implement SFM than to be sure to harvest the granted areas at least once before the whims of a government can give the contract to some other company or allocate the land to some other use.

This is of course not to say that all management plans should be exclusively prepared by public officials. Companies generally have better knowledge and competences than state agencies (more on this below) and indeed, since management plans have become a legal obligation, there have been many positive outcomes for the implementation of SFM. For example, the most progressive companies have embedded into their structure 'management units' and have hired entire new teams to fill those units, which has led to improved knowledge of the resource and more informed decisions taken by company managers, especially in certified concessions (Cerutti et al., 2017). Yet government oversight, forest law enforcement and governance – or, on the demand side, much stronger consumers' awareness than the one elicited today – must improve on par with an increase in the delegation of responsibilities, else only a minority of companies will prepare and effectively implement sound management plans. For the rest of the companies, it will be business as usual.

Lack of clear and respected land-use zoning and the complete delegation of responsibilities to logging companies without concurrent improved forest governance introduce a series of challenges to the implementation of SFM which are worth discussing in more detail. This is done in the next section. We believe such challenges (i) provide a different perspective under which the oft-quoted statistics presented in Table 2 should be read, but also (ii) should stimulate discussions about the meaning of SFM and the potential new paradigms, research and actions needed for its implementation over an area – we often tend to forget – of about 298 million ha, that is larger than the entire European Union, of which we know very little and to which we nonetheless continue to apply the basic tenets of forest management developed a few centuries ago for very different European forests (Nasi and Frost, 2009).

5 From 'nesting dolls' to improved policies

The conditions under which SFM is being implemented in the Congo basin create various 'problems' which impact not only the capacity of governments to effectively manage the forest of which they proclaim themselves owners or custodians, but also the types of social, environmental and financial investments and efforts that private companies and individuals may be willing to deploy on the ground.

First, and most importantly, is a 'jigsaw' problem, or how to draw over the long term an overall sustainable picture of the forest while the building blocks of that picture remain unintelligible to the 'painter'. In fact, given the lack of coherent national plans and coordination mechanisms among various ministries, the current approach is like asking each logging company to draw a piece of a jigsaw puzzle in the hope that – when rotation periods are completed – governments will be able to put all pieces back together to form a sustainable picture. This will clearly not be the case and although such caricature may seem extreme, it does raise relevant questions for current approaches based on often larger-than-concession land units, such as landscapes, zero-deforestation jurisdictions or processes such as national carbon accounting.

Second is the 'missing pieces' problem, or how the long-term picture will look like when various expected but missing building blocks will be accounted for. Given the cruel lack of enforcement, although all legal frameworks mandate the preparation of management plans, in 2018 only about 47% or about 24 million ha of the total area granted as logging concessions had an approved management plan (Table 2). If the DRC's performance in the last couple of years – with about 10 plans approved or 7% of total area – may be considered with optimism, it remains difficult to understand while some companies have maintained (and kept harvesting for many years) their concessions without a management plan ever being prepared and/or approved.

Third, is the '(Russian) nesting dolls' problem, or how the long-term picture will look like when various building blocks will deliver their unsupervised (management) results. As time goes by, a growing number of concessions will enter their second or third rotation, which often means that the most-valued species will get exhausted. Given the historically low level specialisation and further processing – the vast majority of exports is still made of raw logs and unprocessed sawnwood (www.observatoire-comifac.net) – one can expect many companies, notably those least committed to SFM, to abandon their concessions instead of investing in innovative solutions based on today's secondary species. Once the concession is abandoned, it is reattributed. The management delegation to companies however means that the new management plan will reflect the business model of the newcomer, which will again focus on a handful of species, possibly different from those already exploited by the previous company. In summary, each company is thus free to peel away its own layer of the nesting dolls, with the risk that the final drawing on each building block will look extremely different from the one expected under SFM conditions.

These problems create a series of risks and uncertainties on the long-term situation of SFM in the Congo basin's forests. Of course, the argument here is not one about the lack of difficulties and uncertainties about implementing SFM. That remains a very difficult task, and one which requires constant trial-and-errors processes and feedback loops, especially when applied to tropical forests of which we still know very little (Nasi and Frost, 2009). But the existence of risks and uncertainties is no good argument to avoid trying altogether and instead continue with the current situation, especially on the part of governments who put all the good principles – including the precautionary one – in their forest legal frameworks, only to forget them when it comes to implementation and monitoring.

Although criticism of third-party, independent, market-oriented governance systems remain, some of the above risks and uncertainties have arguably been reduced over the last fifteen years in the few patches of forests where management plans are better monitored or in those forests that have been granted and are able to maintain voluntary forest certification schemes (Cerutti et al., 2017). Yet those areas covered a small percentage of the total, or about 4.8 million ha by 2018 (Table 2), which is largely insufficient for governments to be able to monitor and, most importantly, manage and act upon the forests they are mandated to protect. At best, the resulting jigsaw puzzle that governments are able to scrutinise over the national territory is blurred; in most cases, it is entirely unreadable.

One may argue, as it is often the case when it comes to analysing the causes of implementation failures, that in the countries of the Congo basin a preponderant role is played by lack of means and capacities. This may be a good factual explanation of the problem – and one which will be discussed in more detail in the next section – but it is more often used as a bad justification for processes not launched, decisions not taken, and actions not started. The example of Cameroon and the various logging titles that can be granted to industrial logging companies, ultimately leading to unfair competition between supposedly sustainable vs. non-sustainable titles, illustrates this situation very well.

In its various planning strategies for the attribution of logging titles (MINEF, 1999, 2004), the Ministry of Forests reiterates that logging titles such as sales of standing volumes (SSV), which are short-term titles (3 years maximum), generally based in the non-permanent forest domain and thus without SFM mandate, will decrease in number and in timber produced (down to nihil over the course of a few years), in line with the 'priority given to harvesting [through concessions] in the forest domains where forest management is mandated' (MINEF, 1999, p. 9). In reading those documents, it is clear that the government realises the potential for unfair competition between timber harvested in concessions and in SSVs. They harvest similar species, produce the same products (i.e. logs and sawnwood), pay similar taxes and serve similar international markets. SSVs however require no investment in SFM (e.g. no management inventory, no management plan, no long-term social engagement etc.).

In fact, while in 1998 there were 130 active SSVs, as of 2018, about 137 SSVs covering an annual allowable cut of about 137 000 ha were active. Comparatively, the total annual allowable cut in all logging concessions in 2018 was about 201 000 ha. Although discussing the political economy of the granting of SSV is outside the scope of this document, this is a clear example where means, capacities and, most importantly, knowledge about what is 'good' for sustainability are not lacking. Yet, what value proposition is the government making to those companies willing to invest in SFM? The answer seems to be: None.

It is worth mentioning that this is not a criticism of the sovereign decisions that any government remains free to take: The government remains free to grant as many SSVs as it pleases to. It is however a call for governments to adopt decisions for which the means, capacities and background knowledge have indeed existed for decades and still exist today. In this particular case, for example, adopting financial schemes that incentivise companies opting for SFM and disincentivise those going for SSVs seem pretty straightforward options aimed at levelling the playing field in the forest sector to increase SFM's chances of serious implementation. Else, it is easy to do the math for anyone interested in understanding why even those companies willing to adopt and implement SFM may instead prefer the unsustainable, short-term option.

Broadly speaking, this is a clear example whereby, unless a government agrees that SFM must be planned and nourished over the long-term, and the legislator rectifies through regulations the short- vs. long-term divide, SFM will continue to remain the choice of an elite group of companies instead of the mainstream *modus operandi*.

One may argue that Cameroon is the exception among the countries of the Congo basin. Indeed, SSVs do not have analogous titles in other legal frameworks. Yet, we believe this is more a matter of semantics than of reality on the ground, as similar unfair conditions exist in most countries. They are created both through the allocation of non-sustainable titles (such as variously defined 'special titles') which eventually compete with sustainable ones (e.g. Global Witness, 2013 for examples from Cameroon, DRC, Ghana and Liberia), and by allowing concessions to operate in the same country as if different legal frameworks applied, as is the case in various countries where many concessionaires have been allowed to operate without management plans for years (e.g. Southern part of the Republic of Congo, where 77% of logging concessions is still with no plans).

These considerations try to bring to light some of the factors that, when it comes to logging concessions and forest management plans in relation to the implementation of SFM, need to be brought at the table of discussions if an improved situation is our ultimate goal. Yet they remain very much the domain of 'foresters', and only rarely they are considered within a broader perspective, one that considers that these (mostly technical) discussions do not, or should not, occur in a vacuum, as they are part of a broader set of considerations stemming from processes, institutions and variables that generally sit outside the standard domain of those 'foresters'. Those broader (still partly technical, but also eminently political) considerations are the focus of the next section.

6 Improving processes and institutions

Over the course of recent decades, the forest sector in the countries of the Congo has often been the focus of various processes and reform agendas, some specifically targeting forests and the environment, others being broader in nature. In a way, the technical problems highlighted in previous sections are the partial results of such processes, which have been morphing and changing names over the years, but which remain very important when it comes to inventing a new approach to SFM in the Congo basin.

Among the former, the TFAP launched in the 1980s and the International Tropical Timber Organisation (ITTO) aimed at tackling the 'crisis of tropical deforestation' (Winterbottom, 1995, p. 60) and fostering the sustainable use and trade of tropical forests. Among the latter, the 1970s global energy crises resulted in many tropical countries being 'nearly drowned in a sea of debts' (Johnson and Wilson, 1982, p. 211), eventually leading to the signature of a series of SAP with international lending institutions. In some countries, SAPs lasted for more than two decades and came with lending conditionalities which included reforms imposed on the forest sector, forest-related institutions and legal and fiscal frameworks.

Over the years, global and forest-related processes and institutions joined forces to foster the SFM agenda. All processes invariably have had promises and pledges about SFM which have been characterised more by delays in implementation and thwarted expectations than by stark measurable successes. In 1990, for example, the ITTO's members adopted the *Year 2000 Objective*: all exported tropical timber should have come from sustainably managed sources by the year 2000. The year 2000 came and went, with only about one percent of tropical exports sourced from sustainably managed forests (Smouts, 2001). Yet the long-term objectives did not fade away, and politically at least, they maintained an agenda-setting role which contributed to follow-up reforms addressing forest management, forest-based industrial development and the fiscal regimes applied to it: Cameroon had no sustainably managed forest by 2000, but the number of management plans that needed to be approved by 2004 by the relevant Ministry of Forests became a specific conditionality introduced by the country's third SAP; In the DRC, a specific forest policy component was required by the World Bank when it re-engaged with the country at the beginning of the 2000s, with the condition that the then draft Forest Code would be submitted to the National Assembly (The World Bank Inspection Panel, 2007).

Those trends and synergies produced progressive waves of reforms which are still ongoing today under different, arguably more encompassing initiatives, such as those tackling illegal logging and promoting forest law enforcement, forest governance, and better timber production and trade (e.g. FLEG and FLEGT), or those aimed at reducing carbon emissions from deforestation and forest degradation (e.g. REDD+ and zero-deforestation pledges). And in a short circle of history, those initiatives have been receiving renewed support in very recent years from a return to SAP-like conditions in many countries in the Congo basin, under the auspices of the Bretton Wood institutions and other major development and technical partners, as a consequence of yet again stagnant economies and poor financial situations.

Ultimately, these processes worked together over the years to embed into the forest regimes of all Congo basin countries the SFM's concepts and ideas that matured through the initial years of the TFAPs, and were codified during and after the Rio Earth Summit in 1992. Chiefly among those concepts remain the ecological, economic and social functions played by the forests and how the implementation of SFM can make them work in synergy across time to produce results that conserve the forest, improve people's livelihoods, and foster economic growth.

With a few distinctions, those functions are well embedded into the legal frameworks that today regulate forest management in the Congo basin, with various interpretations adding up over the years in parallel with international debates. For example, the limited focus on the tree as a source of logs expanded to non-wood forest products and more recently carbon stocks; the technical rules of silviculture joined the broader interests of social forestry, cultural values, and people's role in managing forests; the logging concession model of colonial heritage expanded to include different uses within the same concession (e.g. agricultural set-asides or riparian zones), as well as new models altogether, such as community and council forests; and public policies started to build synergies with market-driven ones, such as voluntary forest certification schemes (Lambin et al., 2014; Wiersum, 1995).

In turn, those concepts were channelled into national discourses and eventually permeated into the official mandates and missions of new national and regional institutions. For example, many countries moved from Forestry Departments working under the control of Ministries of Agriculture and limited to collecting production and trade statistics, to Ministries of Forests with better-defined, specific mandates to manage, maintain and increase the countries' natural capital. In addition, since the beginning of the 2000s, the Inter-ministerial Commission of Central African Forests (COMIFAC) and the Congo Basin Forests Partnership (CBFP), chiefly among others, work in close relationship to promote the conservation and sustainable management of the Congo basin's forest ecosystems and streamline the efforts of various governments and partners (Bezerra et al., 2018).

Notwithstanding what could indeed be considered as good progress in terms of processes, institutions, and legal frameworks, a huge gap remains between what is *imagined* to happen and what *actually* happens in and around forested areas. This has been one constant concern over the years, which has united State officials, technical partners, private sector, lending institutions, academia, and civil society alike, and which has produced some of the technical 'problems' discussed in previous sections. Arguably, the fundamental reasons for this go well beyond the scope of this document and are to be researched in the political history of the region, including on issues as broad as State formation and extension of sovereignty to vast swathes of land, which remains largely incomplete to this day (for a good overview and more references, see Herbst, 2000). Yet it is worth discussing some key issues nonetheless, in order to be able to link the technical shortcomings discussed above with broader shortcomings, both ultimately hampering the implementation of SFM on a broad scale in the Congo basin.

7 From theory to practice

The issue of SFM impacts on the ground has generally been tackled by two major streams of discussions, literature, and interests which, as is the case for several other topics in nature conservation (e.g. see Sheil, 2016), have been brought forward by foresters on one side, and social and political scientists and practitioners on the other. The two have tended to develop in parallel with very rare interactions: Technical debates on one side, focussing on silvicultural activities occurring within the delimited geographical spaces of logging concessions, on topics as varied as minimum cutting diameters, recovery rates and yields (Putz et al., 2012), processing rates of sawmills, road construction, planning and recovery (Kleinschroth et al., 2017), or, more recently, reduced impact logging for climate (Umunay et al., 2019; Ellis et al., 2019); and political-economic debates on the other side, generally focussing on broader geographies than those represented by concessions, on topics as various as land tenure, civil society participation, private sector engagement, communities' role in, and challenges with, forest management (e.g. Wodschow et al., 2016; Cerutti et al., 2016), informal timber markets (e.g. Lescuyer et al., 2013), state vs. market interventions and many others.

Technically, existing numbers seem to indicate upward trendlines over the course of the past two or three decades. There are more forests managed with officially approved plans, basically from close to zero at the beginning of the 2000s to the figures shown above in Table 2, more communities engaging in forest management, a more vocal civil society, and there is more information available on all of the above than it was the case a few decades ago. Yet, a disconnection remains between where we could arguably be with the implementation of SFM, and where we actually are.

Such a disconnection is, we argue, the elephant in the room which needs to be tackled first and foremost while old initiatives and discourses about SFM morph into new ones and join larger debates about climate change, the expansion of industrial agriculture, landscape approaches, or zero-deforestation pledges in the international political and business agenda.

The reasons for the persistence of this gap in the Congo basin are many as the challenges are financially and politically complex. Conflicting agendas between governments and old and new lending institutions; lack of specific information to 'translate' imported notions about silviculture and SFM into a corpus of locally-relevant knowledge (e.g. Karsenty and Gourlet-Fleury, 2006; Nasi and Frost, 2009); connivance between, and resistance from powerful private interests and corrupt state officials (e.g. Cerutti et al., 2013); and competition between historical (e.g. European) and new (e.g. Asian) players and markets to access the resource, have discouraged investment, innovation and ultimately played against reforms threatening historically established rents and vested interests.

A few considerations and examples illustrate the nature of the disconnect and, at the same time, of the task ahead. First, the number of state officials – very much impacted by the pre-2000s debt crises, devaluations, and follow-up SAPs – tasked with supervising and monitoring forest operations. On *average* across the basin, today each decentralised staff has between 20000 and 30000 ha of logging concessions to monitor.[1] Surfaces have decreased and control has been streamlined and improved in some countries more than others (e.g. Gabon vs. DRC, where logging concessions have increased in recent years without a concurrent increase in supervisory staff), yet overall this ratio has not changed much over the years and across countries (e.g. Global Forest Watch, 2000; Smouts, 2001; CBFP, 2006), and it increases to between 50000 and 100000 ha per staff when all production and conservation forests are considered. Importantly, monitoring is generally supposed to take place with very poor to non-existent budgets and transport facilities. Though big, these averaged numbers do not provide a correct picture of reality on the ground. In fact, most officials are based in capital cities or major towns, while millions of ha of forests remain without management, monitoring or verification activities.

Second, technological innovations such as an increased use of satellite images, GPSs for monitoring and verification purposes, tracking devices, e-government solutions to document validation and so on which could improve management and monitoring tasks – for private companies, civil society, local communities and, most importantly, the State – have indeed been introduced in recent years and – one would hope – even more so in the future. Yet the current human capital – even if it were deployed on the ground where it is needed – is not conducive to the uptake of such technological innovations (de Wasseige et al., 2013; Atyi et al., 2009).

Ultimately, these have so far largely remained *descriptive* tools instead of *management* ones: Maps are more colourful and better-defined, but few people in government systematically exploit them and, most importantly, the information behind them, to monitor and steer what happens in the forests; Management plans are prepared by companies, rubber-stamped by the government, and then largely forgotten in one office or another, instead of becoming *the* instrument through which state officials follow what is happening

---

1 Authors' estimate based on www.observatoire-comifac.org data.

to the natural capital. And while the SFM vocabulary, concepts and technology were continuously expanded and engraved in laws and procedures, the means and curricula of ageing State officials and the few young recruits remained largely focussed on the 1980s' limited notion of timber exploitation. Notions such as Free, Prior and Informed Consent (FPIC), multi-purpose forest inventories, social forestry, community consultations, conflict prevention and management, and many others very much related to modern understandings of SFM, remain very important words into innovative legal frameworks, but useless instruments in the hands of a large number of State officials whose means and budgets are close to nihil and whose knowledge and knowhow are not regularly updated.

For this to change, the number of state officials which can keep up with, understand and implement the global changes about the meaning and objectives of SFM, as well as the more recent technological innovations, must increase exponentially: Financial and political resources, both national and international, must be earmarked for the renovation and improvement of curricula across the board of current ministries and institutions in charge of implementing SFM.

Positive signs exist: The recent examples of exchange programs among Universities, or special funds deployed by the European Union, the World Bank, the German and French Cooperation, and a myriad of other national and international institutions supporting, and promoting exchanges with local Universities and Technical Institutes point into the direction of a generally renewed sense of awareness about this particular issue. But more is needed and a sense of urgency remains, especially in light of current approaches and discourses targeting larger territorial units such as landscapes. Concessions today cover only part of the Congo basin's tropical forests that should be sustainably managed (including sustainable conservation). And yet, any definition of landscape or multi-purpose production encompasses notions of multifunctionality of forests and land which go well beyond the limited means, curricula and mandates of single ministries, be them the forest, environment, or agriculture ones.

8 Conclusions

Towards the end of the last century, Wiersum (1995) assessed 200+ years of experiences with sustainability in forestry and concluded that three general lessons could be drawn. The first two lessons deal with the need to get a better understanding of the different nature of ecological, social and financial dynamics when people and forest resources are considered. The third lesson – the most important for the purpose of this document and in light of the topics discussed here – concerns the 'significance of operational experiences in trying to attain sustainability within a concrete context' (Wiersum, 1995, p. 321).

The recent history in the adoption and implementation of SFM in the Congo basin shows that while the most modern ideas and concepts have shaped the current forest legal framework and are indeed embedded in laws, decrees and regulations, their effective translation into acts on the ground remains wanting. Operational experiences have not been lacking which have led to an embryonic locally-owned understanding of tropical SFM.[2] Yet, the local institutions, knowhow, capacities and locally-based science that can support them are generally not in place. Here and there, very promising experiences have been led by (mostly Northern) Universities and private companies interested in adapting the general models proposed by the theory of SFM to the ecological, social and financial realities of their concessions. But lacking national or regional institutions which can take up that private, localised, knowledge and test it over larger swathes of forests or under different conditions, promising experiences have not yet grown into something transformative.

This document has discussed several issues that speak to the role – or lack thereof – that the State could have in the implementation of SFM. By reflection though, those issues also speak to the role that private companies may assume in such endeavour and to how they may react to the business climate's prevailing conditions. In particular for SFM, the notions of long-term investment and engagement must climb on top of the agenda of policy makers, investors, technical and political partners and, indeed, logging companies themselves.

Currently however, for want of coherent and secure land-use plans, in addition to a generalised situation of lack of monitoring and missed opportunities to promote fair competition among companies, the countries of the Congo basin seem unable to elicit a serious and honest effort towards the implementation of SFM from a vast majority of logging companies. Indeed, existing evidence seems to point in the direction of the State having given up its role of manager, supervisor and verifier of its natural capital.

Ultimately, one can *read* and *hear* a lot more today about SFM in the Congo basin than it was the case two decades ago, especially in multilateral or bilateral meetings with an international audience. Yet on the ground the engagement remains nominal at best and often restricted within the small number of companies which have invested in serious inventories and often adopted third-party certification schemes.

---

2 The interested reader may want to check the 20 documents published in 1998 in the *Série FORAFRI*, also available at: https://www.institutfrancais-gabon.com/catalogue/index.php?lvl=serie_see&id=156. The interested and adventurous reader may want to read the thousands of pages published in the twentieth century in various reviews, diaries, manuals and reports today scattered in forgotten archives across the Congo basin, such as those found in the library of the Yangambi research station, Democratic Republic of Congo (https://forestsnews.cifor.org /56598/decaying-belgian-congo-era-maps-reveal-secrets-of-valuable-african-tree?fnl=en).

Two fundamental consequences of this situation are worth mentioning. First, a generalised lack of well-made, detailed forest information (e.g. inventories), which States in the Congo basin desperately need to plan their future. Second, a lack of serious engagement between the contractual manager of the forest and the populations living in, around and off the forest. All in all, short-termism becomes the norm and it indeed is the rule of the game, as only a handful of committed companies will spend time and resources to establish the solid institutions and frameworks of dialogue requested by the tenets of SFM. This risks not only hampering the implementation of SFM but also rendering null and void the fundamental promise contained in all the forest legal frameworks of the region, that is that forests and the natural capital they contain will deliver better living conditions to local populations.

This document has focussed on the role that States and private companies should play. This is not to underestimate the role that more vocal and engaged local populations and civil society groups can and have to play in the implementation of SFM. They are of utmost importance. Especially so in the rethinking exercise about SFM that this document advocates. Primarily, a new approach to policy making, data-gathering, monitoring and verification activities is needed, one that considers the forest as a whole and that strives to apply SFM with all its facets holistically. And one that assigns a key role to all interested parties, including local populations, civil society organisations and local technical and scientific institutions which can drive the change.

A sense of urgency should embrace us all. The volumes of timber harvested and the areas touched by logging activities are growing and will keep growing, notably so in relation to demography and national and international demand. If the current situation on the ground does not change quickly, the fate of production forest in the Congo basin will be left in the unchecked hands of whomever is granted access to it. History, notably spatially close history such as the one of Cote d'Ivoire, Ghana or Nigeria, seems to indicate that such hands are more naturally inclined to unsustainably exploit the forest for today's financial profits than to sustainably manage it for tomorrow's generations. Most governments in the Congo basin do not seem to have different types of hands. The examples discussed in this document (e.g. 'special' titles allowed to compete unfairly with managed concession, or the overall 53% of logging concessions that keep harvesting the forest without a legally mandated management plan), are all signs pointing in such direction.

Ultimately thus, this document is a call to an entirely new approach to forest management in the Congo basin. One that plans for regular improvements of knowledge, data, means and capacities; extends such knowledge to ever larger areas through local institutions in collaboration with external partners if needed; and - the most difficult part of all - one that gives a prominent role to accountability, notably accountability of governments vis-à-vis their citizens and this planet.

9 References

Atyi, R. E. A., Devers, D., Wasseige, C. D. and Maisel, F. 2009. State of the forests of Central Africa: regional synthesis. In: De Wassegie, C., Devers, D., De Marcken, P. and Eba'a Atyi, R., Nasi, R. and Mayaux, P. (Eds), *The Forests of the Congo Basin - State of the Forest 2008*. Publications Office of the European Union, Luxembourg.

Bezerra, J. C., Sindt, J. and Giessen, L. 2018. The rational design of regional forest regimes: comparing Amazonian, Central African and Pan-European Forest Cooperation. *International Environmental Agreements* 18, 635-56.

Brunner, J. and Ekoko, F. 2000. *La Réforme de la Politique Forestière au Cameroun: Enjeux, Bilan, Perspectives*. World Resources Institute, Washington, DC.

Carrière, S. 2011. Comment nous nous sommes disputés... (notre forêt) : politiques forestières et développement en Côte-d'Ivoire. In: Fontaine, C. (Ed.), *Des forêts et des hommes*. Institut de Recherche pour le Developpement (IRDI), Marseille, France.

Cbfp. 2006. *Les Forêts du Bassin du Congo - Etat des Forêts 2006*. Congo Basin Forests Partnership, Kinshasa, Democratic Republic of Congo.

Cbfp (Ed.). 2007. *Les Forêts du Bassin du Congo - Etat des Forêts 2006*. Congo Basin Forest Partnership (CBFP), Kinshasa, Democratic Republic of Congo.

Cerutti, P. O., Tacconi, L., Lescuyer, G. and Nasi, R. 2013. Cameroon's hidden harvest: commercial chainsaw logging, corruption and livelihoods. *Society and Natural Resources* 26(5), 539-53. doi:10.1080/08941920.2012.714846.

Cerutti, P. O., Ngouhouo Poufoun, J., Karsenty, A., Eba'a Atyi, R., Nasi, R. and Fomété Nembot, T. 2016. The technical and political challenges of the industrial forest sector in Cameroon. *International Forestry Review* 18, 26-39.

Cerutti, P. O., Lescuyer, G., Tacconi, L., Eba'a Atyi, R., Essiane, E., Nasi, R., Tabi Eckebil, P. P. and Tsanga, R. 2017 Social impacts of the Forest Stewardship Council certification in the Congo basin. *International Forestry Review* 19(4), 50-63. doi:10.1505/146554817822295920.

Comifac. 2004. *Plan de Convergence pour la conservation et la gestion durable des écosystèmes forestiers d'Afrique Centrale*. Commission des Forêts d'Afrique Centrale, Yaoundé, Cameroun.

Conigliani, C., Cuffaro, N. and D'Agostino, G. 2018. Large-scale land investments and forests in Africa. *Land Use Policy* 75, 651-60. doi:10.1016/j.landusepol.2018.02.005.

Côté, S. 1993. *Plan de zonage du Cameroun forestier méridional*. Agence Canadienne de Développement International (CIDA) - Ministère de l'Environnement et des Forêt (MINEF), Yaoundé, Cameroun.

Debroux, L., Topa, G., Kaimowitz, D., Karsenty, A. and Hart, T. 2007. *Forests in Post-Conflict Democratic Republic of Congo*. Center for International Forestry Research (CIFOR), Bogor, Indonesia.

De Wasseige, C., Flynn, J., Louppe, D., Hiol Hiol, F. and Mayaux, P. (Eds). 2013. *The Forests of the Congo Basin - State of the Forest 2013*. Weyrich, Belgium.

Ekoko, F. 2000. Balancing politics, economics and conservation: the case of the Cameroon forestry law reform. *Development and Change* 31(1), 131-54. doi:10.1111/1467-7660.00149.

Ellis, P., Gopalakrishna, T., Goodman, R., Putz, F., Roopsind, A., Umunay, P., Zalman, J., Ellis, E., Mo, K., Gregoire, T. and Griscom, B. 2019. Reduced-impact logging for climate change mitigation (RIL-C) can halve selective logging emissions from tropical forests. *Forest Ecology and Management* 438, 255-66.

Global Forest Watch. 2000. *An Overview of Logging in Cameroon*. World Resources Institute (WRI), Washington DC.

Global Witness. 2013. *Logging in the Shadows: How Vested Interests Abuse Shadow Permits to Evade Forest Sector Reforms*. Global Witness (GW), London.

Global Witness. 2015. *Blood Timber – How Europe Helped Fund War in the Central African Republic*. Global Witness, London, UK.

Greenpeace International and The Oakland Institute. 2013. *Herakles Exposed: the Truth behind Herakles Farms False Promised in Cameroon*. Greenpeace International and The Oakland Institute, Oakland, CA.

Gros, J.-G. 1995. The hard lessons of Cameroon. *Journal of Democracy* 6(3), 112-27. doi:10.1353/jod.1995.0048.

Hansen, C. P., Damnyag, L., Obiri, B. D. and Carlsen, K. 2012. Revisiting illegal logging and the size of the domestic timber market: the case of Ghana. *International Forestry Review* 14(1), 39-49. doi:10.1505/146554812799973181.

Herbst, J. 2000. *States and Power in Africa: Comparative Lessons in Authority and Control*. Princeton University Press, Princeton, NJ.

Johnson, W. R. and Wilson III, E. J. 1982. The "oil crises" and African economies: oil wave on a tidal flood of industrial price inflation. *Daedalus* 111, 211-41.

Karsenty, A. 2006. L'impact des réformes dans le secteur forestier en Afrique Centrale. In: Nasi, R., Nguinguiri, J.-C. and Ezzine De Blas, D. (Eds), *Exploitation et gestion durable des forêts en Afrique Centrale*. L'Harmattan, Paris, France.

Karsenty, A. 2018. Commentaire pour CAFI sur la « Note au COPIL – Restitution des travaux de la Commission Technique ad hoc » du Comité Technique de suivi et d'évaluation des Réformes, du Ministère des Finances de la RDC. CIRAD (Unpublished).

Karsenty, A. and Gourlet-Fleury, S. 2006. Assessing sustainability of logging practices in the Congo Basin's managed forests: the issue of commercial species recovery. *Ecology and Society* 11(1), 26. doi:10.5751/ES-01668-110126.

Kleinschroth, F., Healey, J. R., Gourlet-Fleury, S., Mortier, F. and Stoica, R. S. 2017. Effects of logging on roadless space in intact forest landscapes of the Congo Basin. *Conservation Biology: the Journal of the Society for Conservation Biology* 31(2), 469-80. doi:10.1111/cobi.12815.

Lambin, E. F., Meyfroidt, P., Rueda, X., Blackman, A., Börner, J., Cerutti, P. O., Dietsch, T., Jungmann, L., Lamarque, P., Lister, J., Walker, N. F. and Wunder, S. 2014. Effectiveness and synergies of policy instruments for land use governance in tropical regions. *Global Environmental Change* 28, 129-40. doi:10.1016/j. gloenvcha.2014.06.007.

Lescuyer, G., Cerutti, P. O. and Robiglio, V. 2013. Artisanal chainsaw milling to support decentralized management of timber in Central Africa? An analysis through the theory of access. *Forest Policy and Economics* 32, 68-77. doi:10.1016/j. forpol.2013.02.010.

Minef. 1995. *Organisation des forêts de production du Cameroun Meridional – Monographie des forêts domaniales de production et des Unites Forestieres d'Amenagement*. Ministère de l'Environnement et des Forêts (MINEF), Yaoundé, Cameroun.

Minef. 1999. *Planification de l'attribution des titres d'exploitation forestière*. Ministère de l'Environnement et des Forêts (MINEF), Yaoundé, Cameroun.

Minef. 2004. *Planification de l'attribution des titres d'exploitation forestiere*. Ministère de l'Environnement et des Forêts (MINEF), Yaoundé, Cameroun.

Nasi, R. and Frost, P. G. H. 2009. Sustainable forest management in the tropics: is everything in order but the patient still dying? *Ecology and Society* 14(2), 40. doi:10 .5751/ES-03283-140240.

Nasi, R., Cassagne, B. and Billand, A. 2006. Forest management in Central Africa: where are we? *International Forestry Review* 8(1), 14–20. doi:10.1505/ifor.8.1.14.

Ordway, E. M., Asner, G. P. and Lambin, E. F. 2017a. Deforestation risk due to commodity crop expansion in sub-Saharan Africa. *Environmental Research Letters* 12(4). doi:10.1088/1748-9326/aa6509.

Ordway, E. M., Naylor, R. L., Nkongho, R. N. and Lambin, E. F. 2017b. Oil palm expansion in Cameroon: insights into sustainability opportunities and challenges in Africa. *Global Environmental Change* 47, 190–200. doi:10.1016/j.gloenvcha.2017.10.009.

Putz, F. E., Zuidema, P. A., Synnott, T., Peña-Claros, M., Pinard, M. A., Sheil, D., Vanclay, J. K., Sist, P., Gourlet-Fleury, S., Griscom, B., Palmer, J. and Zagt, R. 2012. Sustaining conservation values in selectively logged tropical forests: the attained and the attainable. *Conservation Letters* 5(4), 296–303. doi:10.1111/j.1755-263X.2012.00242.x.

Republic of Cameroon. 1994. *Loi No 94/01 du 20 janvier 1994 portant régime des forêts, de la faune et de la pêche*. Republic of Cameroon.

Republic of Cameroon. 1995. *Décret No 95-53-PM du 23 août 1995 fixant les modalités d'application du régime des forêts*. Republic of Cameroon.

Rights and Resources Initiative. 2015. *Who Owns the World's Land? A Global Baseline of Formally Recognized Indigenous and Community Land Rights*. RRI, Washington DC.

SFIL. 2015. *Situation de nos certificats FSC. Le Directeur Général a tous les partenaires et parties prenantes de la SFIL*. Société Forestière et Industrielle de la Lokoundjé (SFIL), Douala, Cameroon.

Sheil, D. 2016. *Exploring Local Perspectives and Preferences in Forest Landscapes: Towards Democratic Conservation. Tropical Forest Conservation: Long-Term Processes of Human Evolution, Cultural Adaptations and Consumption Patterns*. United Nations Educational, Scientific and Cultural Organization (UNESCO), Paris.

Smouts, M.-C. 2001. *Forêts tropicales, jungle internationale – Les revers d'une écopolitique mondiale*. Presses de Science Po, Paris.

The World Bank Inspection Panel. 2007. Investigation report – Democratic Republic of Congo: Transitional support for economic recovery grant (TSERO) (IDA Grant No. H 1920-DRC) and Emergency Economic and Social Reunification Support Project (EESRSP) (Credit No. 3824-DRC and Grant No. H 064-DRC). Report No. 40746 – ZR. The World Bank, Washington DC.

Umunay, P. M., Gregoire, T. G., Gopalakrishna, T., Ellis, P. W. and Putz, F. E. 2019. Selective logging emissions and potential emission reductions from reduced-impact logging in the Congo Basin. *Forest Ecology and Management* 437, 360–71.

UN-DESA. 2018. *World Population Prospects: The 2017 Revision*. DVD Edition. United Nations, Department of Economic and Social Affairs (UNDESA), Population Division.

Vermeulen, C. and Karsenty, A. 2017. Towards a community-based concession model in the DRC. *International Forestry Review* 19(4), 80–6. doi:10.1505/146554817822295894.

Wiersum, K. F. 1995. 200 years of sustainability in forestry: lessons from history. *Environmental Management* 19(3), 321–9. doi:10.1007/BF02471975.

Winterbottom, R. 1995. The tropical forestry action plan: is it working? *NAPA Bulletin* 15(1), 60–70. doi:10.1525/napa.1995.15.1.60.

Wodschow, A., Nathan, I. and Cerutti, P. O. 2016. Participation, public policy-making, and legitimacy in the EU Voluntary Partnership Agreement process: the Cameroon case. *Forest Policy and Economics* 63, 1–10. doi:10.1016/j.forpol.2015.12.001.

Chapter 5

Trade-offs between management and conservation for the provision of ecosystem services in the southern Patagonian forests

Yamina Micaela Rosas, Laboratorio de Recursos Agroforestales, Centro Austral de Investigaciones Científicas (CADIC), Consejo Nacional de Investigaciones Científicas y Técnicas (CONICET), Argentina; Pablo Luis Peri and Héctor Bahamonde, Instituto Nacional de Tecnología Agropecuaria (INTA), Universidad Nacional de la Patagonia Austral (UNPA), Consejo Nacional de Investigaciones Científicas y Técnicas (CONICET), Argentina; Juan Manuel Cellini and Marcelo Daniel Barrera, Universidad Nacional de la Plata (UNLP), Argentina; and Alejandro Huertas Herrera, María Vanessa Lencinas and Guillermo Martínez Pastur, Laboratorio de Recursos Agroforestales, Centro Austral de Investigaciones Científicas (CADIC), Consejo Nacional de Investigaciones Científicas y Técnicas (CONICET), Argentina

1 Introduction

The concept of ecosystem services (ES) refers to different goods and benefits that society obtains from natural ecosystems (Daily, 1997). One way to graphically represent ES is with a cascade model (Haines-Young and Potschin, 2010). This methodology allowed us to understand the connections between natural and social systems, and the synergies (positives and negatives) that can limit the provision of ES (Reyers et al., 2013). We proposed a model for forested landscapes in southern Patagonia (Fig. 1) that provided an understanding of how different vegetation types (e.g. forests, grasslands, high mountain vegetation) provided specific ecological functions (e.g. biomass growth, water regulation) and produced different ES (e.g. raw material, livestock, water provision). Society

http://dx.doi.org/10.19103/AS.2019.0057.07

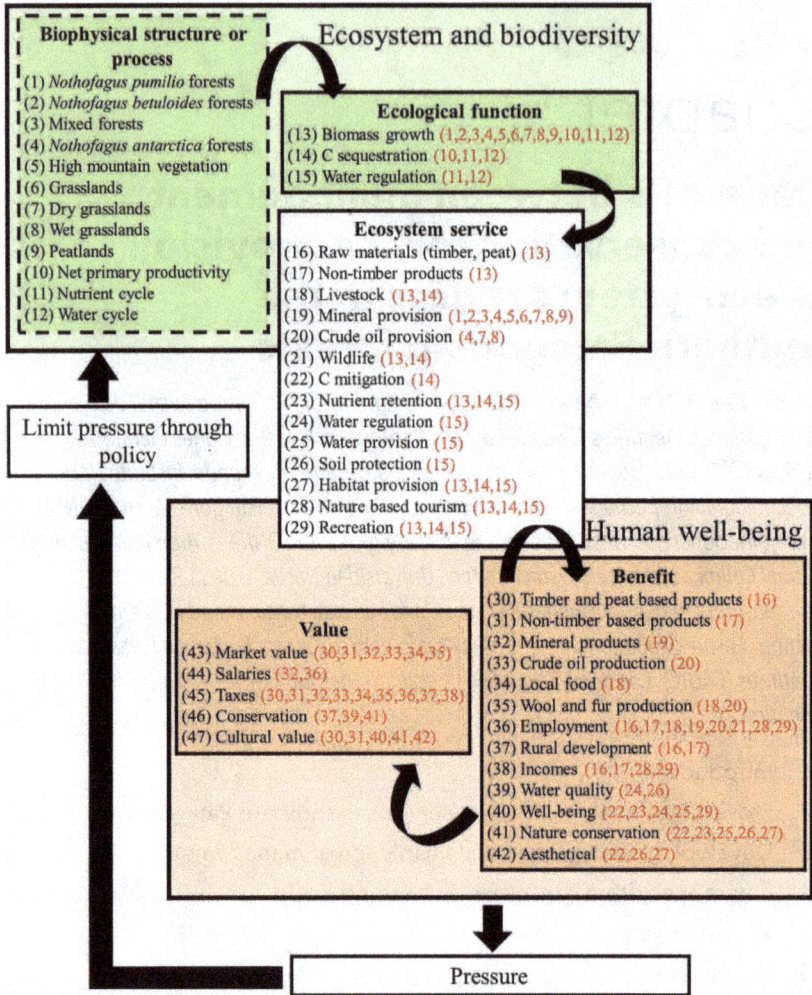

Figure 1 Cascade model of ecosystem services in forested landscapes of southern Patagonia, where red numbers indicate the relationship with the previous component (for details about methodology see Haines-Young and Potschin, 2010).

obtained goods and services from these ES (e.g. timber products, local food, water quality), as well as values (e.g. market value, salaries) for the well-being of individuals and communities.

In particular, forest ecosystems provide critical ES to humanity (FAO, 2010) and play a multifunctional role that balances human needs with the production of other goods and services, including habitat for forest-dependent organisms (Thompson et al., 2011; Lindenmayer and Franklin, 2002). Nevertheless, when ecosystems are managed only for a limited set of goods and services, such as timber from forests, many other ES may be overlooked and therefore undervalued (Thompson et al., 2011; Perera et al., 2018). In forest ecosystems, we can recognize four ES categories (MEA, 2005). For example, some provisioning ES such as timber wood, fiber, or firewood are easy to identify (Gea Izquierdo et al., 2004), while others are less considered such as foods (e.g. fruits, nuts, mushrooms, honey, or spices), pharmaceutical plants, and other non-woody industrial products (Quintas-Soriano et al., 2016; MEA, 2005). In addition, silvopastoral systems (SPS) are designed to increase the provision of ES from managed forests, for example including livestock in the ecosystems (e.g. cattle, goats, sheep) generates more trade products (e.g. meat, milk, wool, leather) (Peri et al., 2016a).

Water is another provisioning ES for economic and social purposes. The benefits of clean water include both direct (e.g. irrigation, crops, or recreation) and indirect uses (e.g. aesthetics or nature conservation). Forests can achieve the requirements for both quantity and quality for different uses, for example by roots reducing sedimentation, filtering nutrients, and retaining and storing water to avoid downstream flooding (Kreye et al., 2014). Preserving forest cover can be considered a relatively low cost but effective way to maintain clean water (Ernst et al., 2004), where climatic changes and timber harvesting have significant impacts on water resources by altering eco-hydrological processes (Sun and Vose, 2016).

Forest ecosystems are important in the role of regulating ES; forests contribute to climate stability by removing greenhouse gases and other pollutants from the atmosphere and by filtering dust particles from the air. Changes in forest structure from deforestation, degradation, and land-use change directly affect the risk of natural disasters (e.g. storms, floods, and droughts) (FAO, 2010), thereby increasing the vulnerability of natural ecosystems (MEA, 2005). Soil retention and water regulation also depend on maintaining forest cover and intact root systems. Tree roots stabilize the soil and regulate water run-off; foliage intercepts rainfall thus preventing compaction and erosion of bare soil (De Groot et al., 2002; Panagos et al., 2015). Natural pest control is another important regulating ES, where some forest species (e.g. wasps, owls, and bats) help humans to regulate diseases (González et al., 2015; Quintas-Soriano et al., 2016). Pollination also is another crucial regulating ES (e.g. for production of vegetables, legumes, and fruits) provided by species that only live in forested lands (Martins et al., 2015; Quintas-Soriano et al., 2016).

Aesthetic beauty, recreation, and ecotourism are important ES of forests for rest, relaxation, and refreshment (De Groot et al., 2002), including many opportunities for recreational activities (e.g. hiking, camping, and fishing). Forest ecosystems are also used as motivation and sources for cultural and artistic inspiration (e.g. books, film, and architecture) and sense of place for local communities.

Finally, supporting ES are indispensable to maintain the other three ES categories, so their influence on human well-being is indirect (MEA, 2005). Soil formation in forest ecosystems is one of the most important ones, occurring as a very slow process (Lal and Lorenz, 2012). The forest canopy influences nutrient and water cycling, alters hydrological conditions by redirecting precipitation, reduces snow accumulation, and removes soil water through transpiration (Prescott, 2002). These variables affect decomposition of animal and plant organic matter and the release of nutrients, which also are strongly influenced by climate (De Groot et al., 2002). Productivity of ecosystems is another supporting ES that is linked with the provision of several ES. Forest landscapes represent 75% of terrestrial gross primary production and 80% of total plant biomass (Perera et al., 2018). This service is also greatly affected by climatic events (e.g. droughts); increasing temperatures reduce productivity in the rainforests in the Amazon but increase productivity in the Northern Hemisphere (Zhao and Running, 2010; Chen and Luo, 2015). Finally, habitat provision for all wild plants and animals is another supporting ES, which favors the health of habitats, a necessary precondition for the provision of all the other ES, directly or indirectly (De Groot et al., 2002).

1.1 The role of biodiversity to support the delivery of forest ecosystem services in **Nothofagus** *forests*

Biodiversity is assumed to be critical for ES supply (MEA, 2005) and functional processes (Thompson et al., 2011; Mori et al., 2017). In fact, some authors indicate that biodiversity is itself an ecosystem service because it is the basis for nature-based tourism or the regulation of diseases (Mace et al., 2012). Human society has been built on biodiversity (Díaz et al., 2006), strongly compromising the delivery of ES and processes by altering ecosystems especially by deforestation (Baillie et al., 2004; MEA, 2005; Díaz et al., 2006).

Temperate regions of the world have been the most extensively altered by human activities, with significant impacts on the provision of goods and services (Franklin, 1988; Lindenmayer et al., 2012). However, temperate forests in South America represent almost one-third of the world's few remaining undisturbed temperate forests (WRI, 2003). These forests are used primarily for production of wood and non-wood forest products, and they are important for regional economies (Li et al., 2011). In part, these benefits have been made possible by exploiting certain ES through well-established markets (e.g. timber, livestock, crops) to the detriment of ecosystem functions, and the underlying capacity to continue to provide other non-market services (e.g. recreational activities, carbon sequestration, climate regulation, water quality) (MEA, 2005). Non-market benefits of forests are directly linked to biodiversity value and forest habitat quality (Spagarino et al., 2001; Martínez Pastur et al., 2002). When ES are not linked to commercial markets, the chance of degradation is increased (Li et al., 2011). For this, the main difficulty remains to define forest management strategies for non-monetary ES (Perera et al., 2018).

The conceptual framework of the MAES initiative (Mapping and Assessment of Ecosystems and their Services, 2014) may be used to describe the multifaceted role of biodiversity to support the delivery of ES and to assess the status of the *Nothofagus* forest ecosystem in South America (Fig. 2).

Genetic diversity of wild species is involved in the quality and production of provisioning services (e.g. wood volume) influencing the susceptibility to pest and climate variations and global warming (De Groot et al., 2010). *Nothofagus* forests present a low genetic diversity growing in a wide range of environments and ecological conditions. Ramírez et al. (1997) determined that *N. antarctica* (ñire) is the species with the largest phenotypic plasticity. This phenotypic plasticity is related to climatic (temperature, rainfall, and radiation) and topographic (elevation) variables, and can generate different morphotypes (e.g. krummholz; Soliani and Marchelli, 2017). *Nothofagus pumilio* (lenga) and ñire are widely distributed across the Los Andes cordillera (more than 2500 km), which generate few genetic geographic regions (n = 7), due to their homogeneity (Soliani and Marchelli, 2017).

Species richness is highly significant for the provision of economic (e.g. biomass production) and ecological services (e.g. water regulation, soil C sequestration) by forests (Perera et al., 2018). Richness in *Nothofagus* forests is low compared to other temperate forests (Lencinas et al., 2008a; Hudson et al., 2017) with few specialist and many generalist species. Besides this, these forests present several endemic and relict species; due to landscapes that were molded by glaciers (Rabassa et al., 2000) that isolated many species

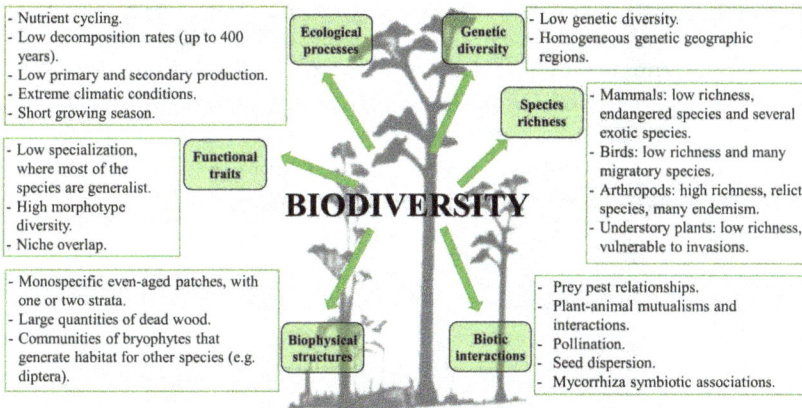

Figure 2 Biodiversity supports the delivery of ecosystem services: characterization of the different multifaceted roles in *Nothofagus* forests.

for thousands of years. For example, few mammal species exist, and most of them have limited habitat distribution (e.g. *Lontra provocax*, *Pudu pudu*) (Bonino, 2005) or are endangered (e.g. *Hippocamelus bissilcus*) (Rosas et al., 2017). Other species, such as *Lama guanicoe* or *Puma concolor*, are widely distributed and occupy most of the ecosystems at the landscape level (Bonino, 2005). Besides this, several exotic species were introduced to enhance the provisioning ES (e.g. hunting and fur production) but significantly modified the ecosystems (e.g. *Castor canadensis*) (Wallem et al., 2010). Most bird species are generalists, and forest birds have very low richness (Lencinas et al., 2005, 2009) with close to 33% migratory species (Ippi et al., 2009; Martínez Pastur et al., 2015). In contrast, arthropods have higher richness assemblage (close to 400 species) but many still are not taxonomically described (Lencinas et al., 2008b) and there are many unique, rare, and relict species (Lanfranco, 1977; McQuillan, 1993). Finally, understory plants in *Nothofagus* forests are very limited compared with other vegetation types, where the species assemblage reaches to 20–40 vascular plants (Lencinas et al., 2008a).

Biotic interactions among species (predation, parasitism, competition, and facilitation) have important consequences for ES and include direct interactions, such as pollination, mycorrhizal fungi, and nitrogen-fixing microorganisms, links between plants, herbivores, and seed dispersers, and organisms that modify habitat conditions such as beavers. Indirect interactions involving more than two species include pathogens that control herbivores and thus avoid overgrazing of plants (MEA, 2005); change in this interaction can lead to disproportionately large, irreversible, and often negative alterations of ecosystem processes. Different species (e.g. insects, wasps, plants) have biotic interactions among them such as prey-pest relationships controlling species that damage timber tree species (González et al., 2015; Quintas-Soriano et al., 2016), or plant-animal and plant-plant mutualisms that provide services (e.g. recruitment facilitation through nursing) (Henríquez and Lusk, 2005). Pollination is another ES supporting the natural forests. Morales and Aizén (2006) found that flora of the temperate forests of Patagonia supports a diverse flower visitor community, composed mostly of native insects and a few abundant alien insects, where the most important order is Hymenoptera. Armesto et al. (1987) identified different plant-animal interactions, with bird and lizard frugivores contributing to seed dispersal. Finally, symbiotic mycorrhizal interactions are crucial for seedling survival in disturbed forest environments (Hewitt et al., 2018).

Biophysical structures provide ecological conditions and habitat to support the biodiversity associated with ES provision (MEA, 2005). *Nothofagus* forests have simple structures, mostly monospecific even-aged patches with one or two strata (Martínez Pastur et al., 2013), and these stand conditions support a relatively low richness compared to other temperate forests (Lencinas et al., 2008a). Besides this, large quantities of woody debris accumulate in the forest floor that supports many insect groups (Spagarino et al., 2001; Grove and Meggs, 2003) and bryophytes (Lencinas et al., 2008c; Müller et al., 2015). When forests are harvested by variable retention methods, plant species richness increases due to an increment in light and water availability; however, native forest insect specialists decrease due to the loss of habitats (e.g. old-growth trees) (Soler et al., 2016). In other management systems, such as silvopasture, the presence of coarse woody debris was identified as a negative synergy with provisioning services related to grass biomass in the understory (i.e., 20% of debris cover reduced the same percentage of grass biomass) (Peri et al., 2016a).

Functional traits reflect adaptations to variation in the physical and biotic environment and trade-offs (ecophysiological and/or evolutionary) and assist in supporting biodiversity associated with ES provision. The type, range, and the relative abundance of functional traits exert a significant control over different ES across a range of organisms and ecosystems (de Bello et al., 2010). Most of the species in *Nothofagus* forests have low specialization, being mainly generalists (Lencinas et al., 2008a; Martínez Pastur et al., 2016b). Many species have high morphotype diversity that develops in different habitats (e.g. from the sea-level to tree-line). Premoli and Brewer (2007) found that lenga had a complex eco-physiological pattern in tree growth along elevation gradients, where the use of water (stomatal conductance) appeared to be more responsive to environment than genetics. Lenga growing at high elevation had low height, wide branch angles, and greater branching ratios than low elevation trees (Premoli et al., 2007). The high percentage of generalist species generated a high niche overlap among them, generating several trade-offs in ES (e.g. *Lama guanicoe* and livestock) (Soler et al., 2012).

Ecological processes of natural ecosystems are essential for biodiversity by regulating the dynamics of ecosystems and the structure and dynamics of biological communities (Mace et al., 2012). *Nothofagus* forests depend on nutrient availability (Gargaglione et al., 2014) but the decomposition rate is low; logs require more than 400 years to fully release immobilized nutrients (Frangi et al., 1997). In high latitudes, as in Tierra del Fuego, the extreme climatic conditions and the short growing season (Massaccesi et al., 2008) lead to low primary and secondary production, reducing potential biodiversity and the associated ES.

Quantifying the relationship among biodiversity and the different levels of ecosystem functions has only been achieved in a few experimental situations and remains an area of active research (MEA, 2005). The relationship between biodiversity and human well-being cannot be determined due to lack of data to support strong conclusions (Cardinale et al., 2012). Global platforms of geo-tagged digital images (e.g. Google Earth) can be useful tools to identify and map cultural ES at regional scales (Casalegno et al., 2013; Martínez Pastur et al., 2016a). In addition, this technique allows identification of social and biophysical features associated with the provision of the cultural ES and allows the assessment of spatial trade-offs and synergies.

Recent methodological advances have improved the assessment of species distributions, synergies, and trade-offs of various ES and biodiversity at different spatiotemporal scales (Raudsepp-Hearne et al., 2010; Martínez Pastur et al., 2017). Andrew et al. (2014) proposed some alternatives: (i) species occurrence maps, which only are useful when the link between species and services is well understood or when the species is the service itself (e.g. food); (ii) forest cover-type maps, which use the spectral data of the dominant canopy or the heterogeneity of reflectance values within a set of pixels and relate them with biodiversity; and (iii) modeling habitat species distributions and biodiversity; this kind of map can be indirectly developed using remotely sensed environmental correlates, and was employed around the world (Guisan and Zimmermann, 2000; Hirzel et al., 2002; Soberón and Peterson, 2005; Elith and Leathwick, 2009) and in Patagonia (Martínez Pastur et al., 2016b; Rosas et al., 2017, 2018). These methods have advantages and disadvantages according to the objectives, data available, and landscape scale (Rodríguez et al., 2007).

1.2 Forest ecosystem services and management strategies in Nothofagus forests

The identification of the biodiversity components that are responsible for ES delivery is crucial to develop management and conservation strategies at different landscape levels. Measuring change in populations is important for understanding the link between biodiversity and ecosystem function, as significant changes in populations may imply changes in the function of ecosystems (MEA, 2005). Negative implications for biodiversity are greatly exacerbated in areas of high human activity (Easterling et al., 2001; Harcourt et al., 2001); mainly this is spatially coincident with forests of high species richness or endemism (Ceballos and Ehrlich, 2002). Therefore, it is relevant to develop sustainable management plans in forest and agro-forest ecosystems considering the substantial contributions of forest ES to global society and the biodiversity that they support (e.g., about three-quarters of terrestrial taxa) (Daily and Ehrlich, 1995; Mori et al., 2017).

Studies in *Nothofagus* forests that analyzed the impact of forest management showed a significant overall biodiversity loss (Deferrari et al., 2001; Spagarino et al., 2001; Martínez Pastur et al., 2002; Ducid et al., 2005). However, when forest diversity assemblages were analyzed at the landscape level, only some insects groups (e.g. coleopteron and dipteran) were greatly affected by forest management (Lencinas et al., 2007, 2008b). The application of adaptive forest management strategies such as variable retention (Gustafsson et al., 2012) is needed to integrate economic, societal values and biodiversity conservation in lenga forests of southern Patagonia (Martínez Pastur et al., 2007, 2009; Lencinas et al., 2007, 2009). Another example is the implementation of SPS in ñire forests that combine trees and grasslands or pastures under grazing in the same unit of land (Peri et al., 2016a). These systems provide diversification of farm incomes, increase provisioning services (e.g. timber and animals), and enhance animal welfare and production indirectly by shelter and biomass forage for livestock. SPS provide advantages compared with conventional grazing and traditional timber production practices because they increase biodiversity conservation and multiple ES (water regulation, carbon sequestration, nutrient cycling) (Fischer et al., 2011; Durán et al., 2014; Peri et al., 2010). Trade-off between forest management and biodiversity has been widely reported, with major implications for environmental management (Cordingley et al., 2016; Martínez Pastur et al., 2017). Understanding the processes behind forest ES provisioning, as well as their trade-offs with biodiversity conservation, is a useful tool to support spatial planning and land management (Li et al., 2011; Carvalho-Santos et al., 2016).

Management and conservation strategies must be designed to apply at different landscape scales, and one alternative is considering land-sparing and land-sharing strategies (Phalan et al., 2011). Land-sharing can be considered as a multifunctional approach to land use, where delivery across multiple ES is prioritized. Introducing habitat heterogeneity and providing refuges for species retain services such as pollination, water quality, and biodiversity for food production (Whittingham, 2011). However, this implies the need for larger areas to be managed to maintain both yield targets and other ES. The other alternative is the land-sparing strategy that spatially segregates land areas for production and areas prioritized for other ES. However, this strategy has shown limitations for protecting species survival at the landscape level (Lindenmayer and Franklin, 2002; Phalan et al., 2011). Nowadays, the tendency is to design a multi-objective management strategy as for example the protection of areas of high species diversity (Rands et al., 2010) and ES such as climate regulation by carbon sequestration (Strassburg et al., 2010). However, new strategies for the protection of multiple ES are likely to be necessary, consistent with the rising popularity of the ecosystem approach to spatial planning and land management (Goldman et al., 2008).

2 Provision of forest ecosystem services in southern Patagonia

Southern Patagonia includes Santa Cruz and Tierra del Fuego provinces (Argentina) (Fig. 3), where native forest is distributed from 46° to 55° S. The different forest types and species were related to rainfall patterns, altitude, temperatures, and soil quality condition (Veblen et al., 1996; Martínez Pastur

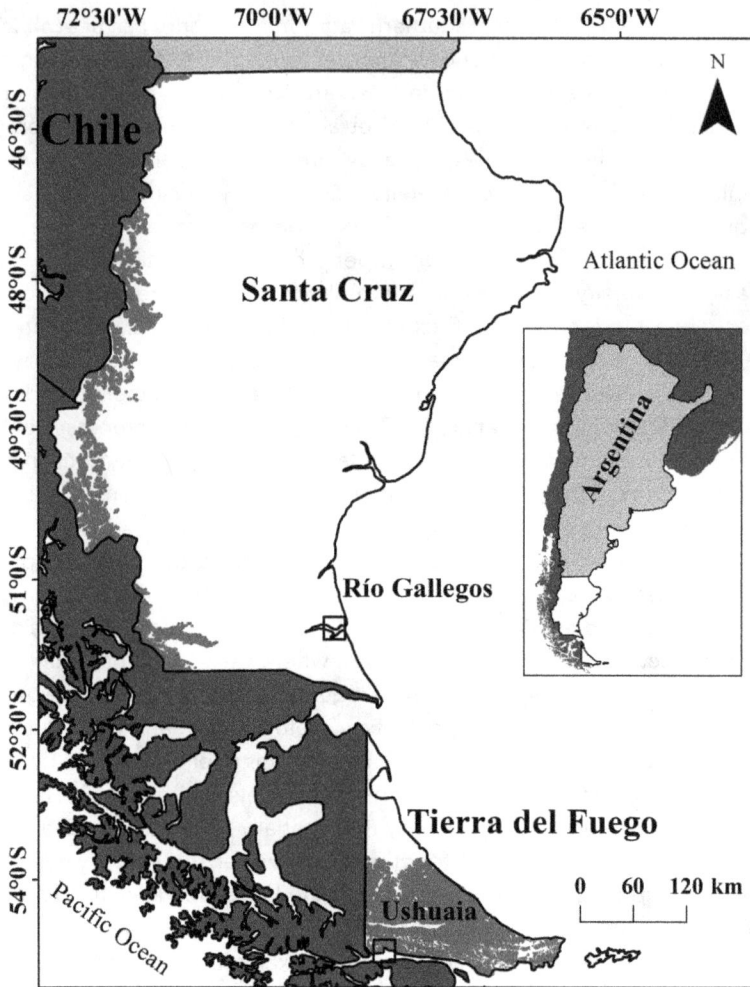

Figure 3 Forest ecosystem (green) in southern Patagonia (Argentina). Squares represent the capital cities.

et al., 2016b; Lencinas et al., 2008a). *Nothofagus* forests in Santa Cruz grow in a narrow area in the Andean Mountains, while in Tierra del Fuego an extended area from the central part of the island to the sea has *Nothofagus* forests. These forests have two deciduous species (lenga and ñire) growing in pure stands, and one evergreen species (*N. betuloides*) growing in mixed forests with other secondary tree species (e.g. *Drimys winteri, Embothrium coccineum,* and *Maytenus boaria*).

ES provided by traditional management practices are clearly identified, which provide timber, firewood, forage, and habitat for cattle (Peri et al., 2016a;

Martinez Pastur et al., 2016b). However, another ES, tourism, has been recently identified as important in the economy of the region (Martinez Pastur et al., 2017). In this region, the relationship between biodiversity and ES is still poorly understood (Martínez Pastur et al., 2017); however, in the recent years there have been some efforts to incorporate the concepts of ES and their trade-off with biodiversity.

Provisioning ES change according to the forest type and the management strategy (Fig. 4). Ñire forests are mainly used for livestock and wood for different uses (e.g. firewood and timber for rural construction) (Peri et al., 2016a). Data from forest inventories and landscape studies (Peri and Ormaechea, 2013) allowed us to characterize the provisioning ES. Our studies showed that 56% of the ñire forests outside the protected areas have lower silvopastoral potential (<500 kg dry weight (DW) ha^{-1} of understory forage for livestock and <100 m^3 ha^{-1} of wood products), 38% have intermediate potential (<500 kg DW ha^{-1} and 100–200 m^3 ha^{-1}), and only 6% have high potential (>500 kg DW ha^{-1} and >200 m^3 ha^{-1}). Lenga forests were mainly used for timber production (Martinez Pastur et al., 2013). Our studies showed that 54% of the lenga forests outside the protected areas are non-timber (<30 m^3 ha^{-1}), while timber production forests had different yield qualities: (i) 19% between 30 and 45 m^3 ha^{-1}; (ii) 24% between 45 and 60 m^3 ha^{-1}; and (iii) only 3% with >60 m^3 ha^{-1} (Fig. 4).

Regulating ES can be characterized through proxies, such as soil nutrient content (Peri et al., 2016b, 2018, 2019) (Fig. 5). Soil organic carbon content (Peri et al., 2018) and soil nitrogen content were estimated through models (n = 150 plots) fitted with more than 40 climatic, topographic, and landscape explanatory variables. The results showed that soil organic carbon content

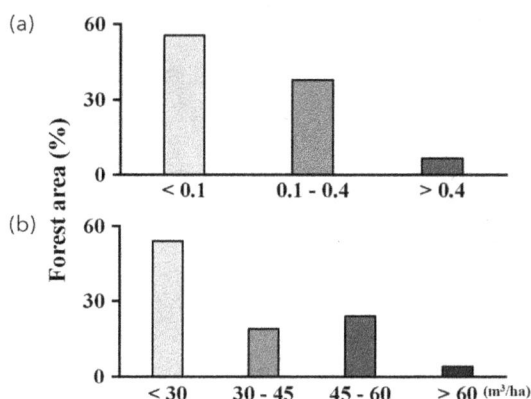

Figure 4 Characterization of provisioning ecosystem services in Santa Cruz forests: (a) silvopastoral potential of *Nothofagus antarctica* forests considering dry weight biomass of the understory and wood products volume (0–1 increasing the provisioning services), and (b) timber volume for sawmill industry of *N. pumilio* forests.

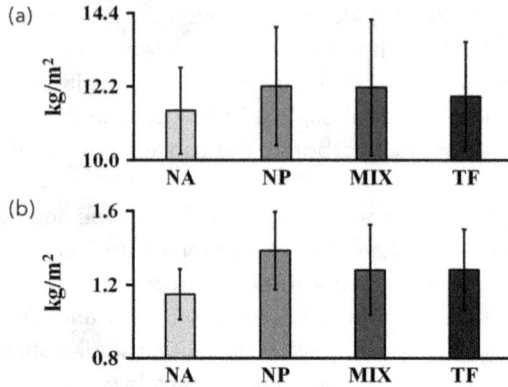

Figure 5 Characterization of regulating ecosystem services in Santa Cruz forests: (a) soil organic carbon content, and (b) soil nitrogen content. NA = *Nothofagus antarctica*, NP = *N. pumilio*, MIX = mixed evergreen forests, TF = total forests. Bars indicate standard deviation.

was higher in forested landscapes (11.9 kg-C m^{-2}) than the average of Santa Cruz province (5.3 kg-C m^{-2}). At the forest level, the higher values were found in lenga and mixed evergreen stands (12.2 kg-C m^{-2}) compared to ñire forests (11.5 kg-C m^{-2}). Soil nitrogen content was higher in lenga forests (1.4 kg-N m^{-2}) than in mixed evergreen forests (1.3 kg-N m^{-2}) or in ñire forests (1.1 kg-N m^{-2}) (Fig. 5).

Supporting ES were characterized through habitat quality and the productivity of the different forest types (Fig. 6). Habitat quality can be related to the degree of impact on the natural environment or nature loss, due to

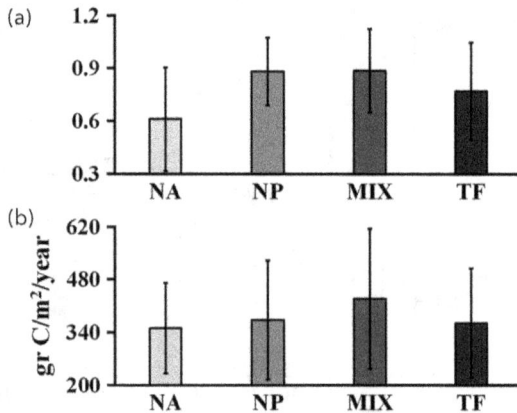

Figure 6 Characterization of supporting ecosystem services in Santa Cruz forests: (a) habitat index (0–1) where 1 represents the lowest human impact, and (b) yearly net primary productivity. NA = *Nothofagus antarctica*, NP = *N. pumilio*, MIX = mixed evergreen forests, TF = total forests. Bars indicate standard deviation.

human influence. One methodology proposes the human footprint index (HFI) (Sanderson et al., 2002), that quantifies the human pressure over the environment, such as (i) urban areas, (ii) transportations routes, (iii) extractive activities, or (iv) land use (e.g. livestock density). We developed the HFI for Santa Cruz province, and used a habitat index of one (1-HFI) to characterize the fully intact forest landscape (Watson et al., 2018). The better habitat quality was found in mixed evergreen (0.89) and lenga forests (0.88), well represented in the protected areas. Ñire forests had lower values (0.61), and they were mainly located in areas with ranching activities. The forests differed in productivity, and we used the yearly net primary productivity as a proxy (Zhao and Running, 2010). The higher values were found in mixed evergreen forests (429.5 gr-C m^{-2}yr^{-1}), followed by lenga (372.6 gr-C m^{-2}yr^{-1}) and ñire forests (351.4 gr-C m^{-2}yr^{-1}).

Cultural ES contribute to different dimensions of human well-being (Vilardy et al., 2011; Russell et al., 2013). These ES have rarely been integrated into a decision-making framework because of their intangibility, complex relationship with biophysical variables, and the difficulty of quantifying their multiple social values (Daniel et al., 2012). New methodological approaches are needed to quantify the social importance of cultural ES and to analyze their spatial patterns (Casalegno et al., 2013). The main challenge is to map cultural ES, especially in those areas with low data availability, such as southern Patagonia. Martínez Pastur et al. (2016a) proposed a methodology based on analyses of photos that local people uploaded to the web. We developed maps of cultural ES (aesthetic, existence, recreation, and local identity) in Santa Cruz province according to the outputs of Martínez Pastur et al. (2016a) into a GIS database using the tool kernel density. Then, we combined them into a single map (standardized values (1 = high, 0 = low) of the average of the four cultural ES maps) (Fig. 7), where higher values indicated that more cultural ES were identified in the area. Higher values were related to cities close to protected areas (e.g. Luis Piedrabuena near to Monte León National Park or Calafate and Chaltén close to Los Glaciares National Park), but also were related to main routes that connect cities and tourist areas. This map showed higher values in mixed evergreen (0.54) than lenga (0.39) and ñire (0.25) forests. The same trend was observed when different cultural ES were considered for each forest type (Fig. 8). Aesthetic values were greater in mixed evergreen than in lenga forests, which grow close to mountains, and both were greater than in ñire forests that grow in lowlands. Existence values follow the same pattern (mixed evergreen > lenga > ñire) due to the occurrence of special species (e.g. *Embothrium coccineum* or *Hippocamelus bisulcus*) and the closeness to emblematic tourist attractions (e.g. Perito Moreno glacier). Recreational values were greater in ñire forests than in mixed evergreen and lenga forests due to better accessibility from cities (e.g. Rio Turbio) and also to the closeness of the national parks.

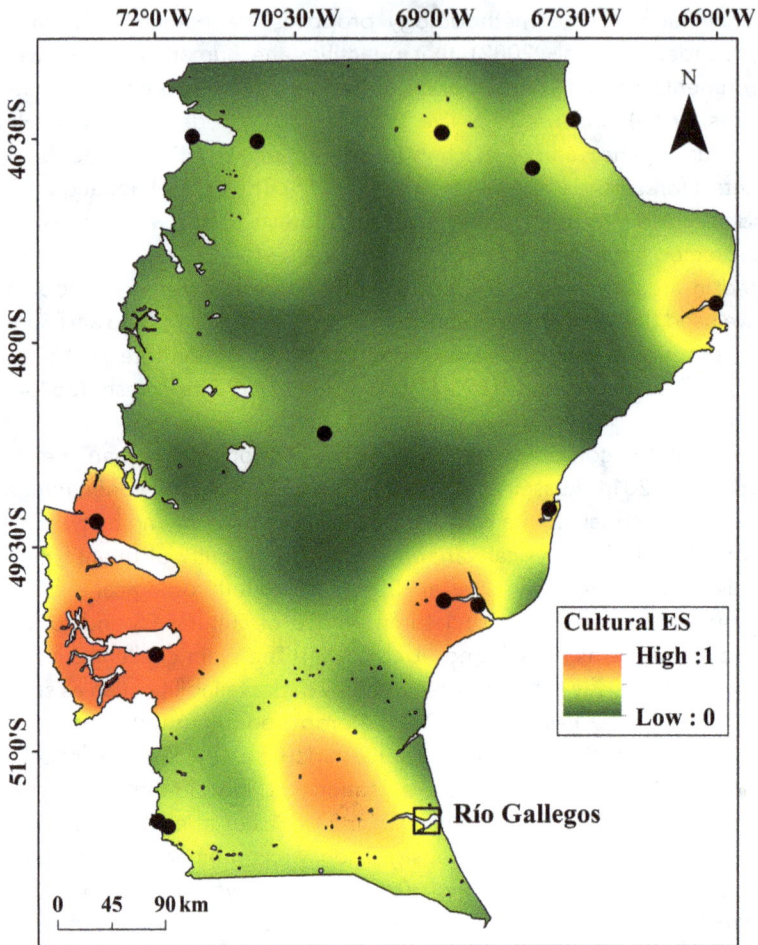

Figure 7 Map of cultural ecosystem services in Santa Cruz province. Each pixel represents the standardized values of the average of aesthetic, existence, recreation, and local identity, according to the outputs of Martínez Pastur et al. (2016a). Black dots represent cities with more than 10 000 inhabitants and the square represents the capital city.

Local identity (e.g. traditional activities) was greater in mixed evergreen than in lenga and ñire forests, and it was related to the most important tourist cities, human settlements, and ranching activities.

Biodiversity plays a major role in the provision of ES (Mace et al., 2012; Loreau and de Mazancourt, 2013; Harrison et al., 2014; Martínez Pastur et al., 2016b), and its valuation characterizes the stability and health of natural ecosystems (Thompson et al., 2011; Mori et al., 2017). However, it is difficult to map biodiversity in areas with low data availability. For this, some methods based on ENFA models (Hirzel et al., 2002) were used to develop

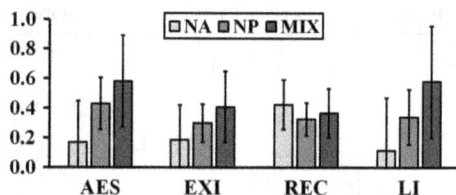

Figure 8 Characterization of cultural ecosystem services (EAS = aesthetic, EXI = existence, REC = recreation, LI = local identity) in different forest types (NA = *Nothofagus antarctica*, NP = *N. pumilio*, MIX = mixed evergreen) in Santa Cruz province. Bars indicate standard deviation.

potential biodiversity maps (PBM) (e.g. PBM for *Nothofagus* forests in Tierra del Fuego, Martínez Pastur et al., 2016b). In Santa Cruz province, potential habitat suitability maps for different species groups were developed: (i) native deer (*Hippocamelus bisulcus*) (Rosas et al., 2017), (ii) lizards (eight species) (Rosas et al., 2018), (iii) darkling beetles (ten species), and (iv) understory plant species (53 species). Here, we combined these maps to develop a unique PBM for *Nothofagus* forests in Santa Cruz province (Fig. 9). Values were classified according to potential (from 1% to 100%), where higher values indicated better habitat quality in the area. We found that potential biodiversity changed through latitudinal and longitudinal gradients (i.e. higher values were found in south than north and east than west), and with forest types. Ñire forests had higher average values of potential biodiversity (65%), followed by lenga (51%) and mixed evergreen forests (45%) (Fig. 10).

Understanding and predicting how multiple ES co-varied with other drivers became relevant for guiding sustainable environmental management for human well-being (Bennett et al., 2009; Raudsepp-Hearne et al., 2010). For this, quantifying multiple ES of forest ecosystems is important, especially because (i) ES are present in each forest type; (ii) to determine the main factors that contribute to the provision of each ES; and (iii) identifying linkages between biodiversity and each ES. Understanding these relationships is crucial for developing management strategies for current and future human well-being (Maskell et al., 2013; Schindler et al., 2014, 2016) and to identify the potential synergies and trade-offs among the ES and biodiversity (Martínez Pastur et al., 2017). With the outputs (Figs. 4-10) we were able to identify the main provisioning ES and potential biodiversity for each forest type. We found the following:

(i) Mixed evergreen forests were the most important for supporting cultural and regulating ES. However, because the provisioning ES and the potential biodiversity were low, this forest type was well represented in the protection areas (79% of the forests were protected). Because of this, there was a low probability of potential trade-offs among the ES provision and biodiversity for this forest type.

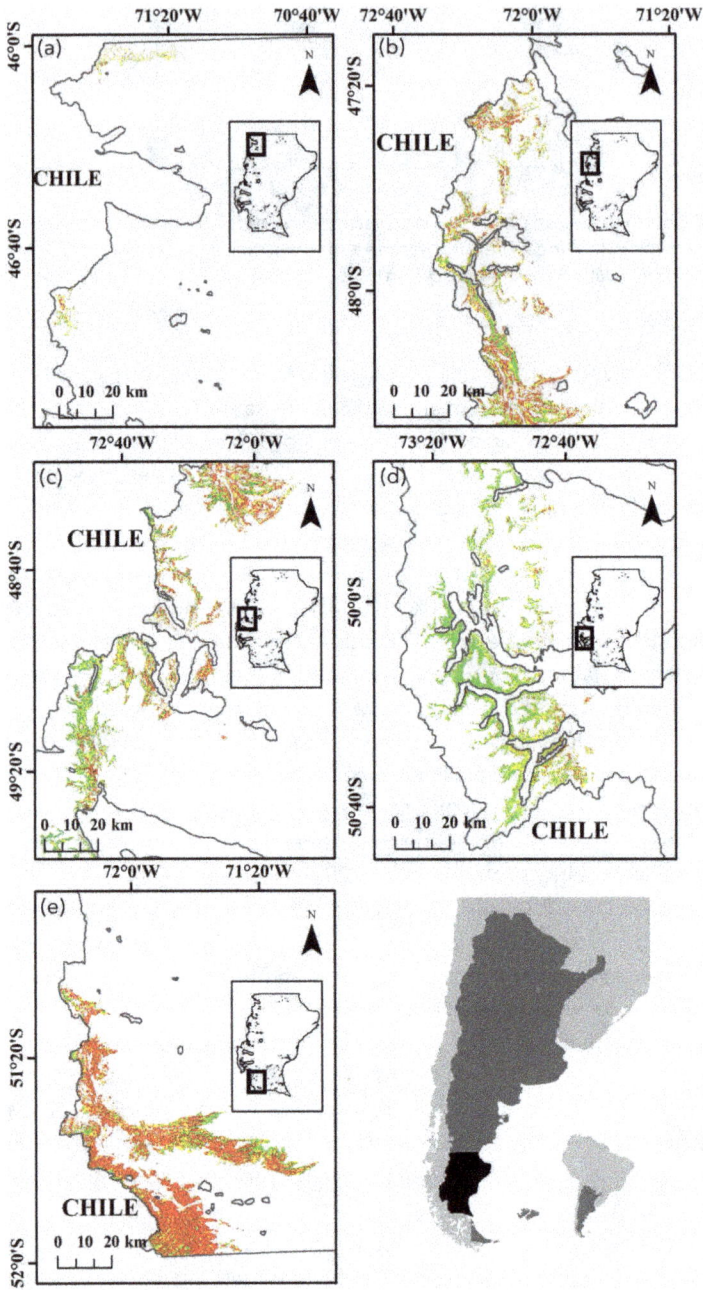

Figure 9 Map of potential biodiversity of Santa Cruz forests. Low potential = green (1–49%), medium potential = yellow (50–63%), and high potential = red (64–100%). (a) Lago Buenos Aires, (b) Lago Pueyrredón, (c) Lago San Martín, (d) Lago Argentino, and (e) Río Turbio.

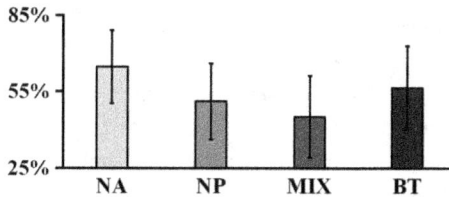

Figure 10 Potential biodiversity of Santa Cruz forests, where the percentage indicates that more species means better habitat quality (see methods in Martínez Pastur et al., 2016b). NA = *Nothofagus antarctica*, NP = *N. pumilio*, MIX = mixed evergreen forests, TF = total forests. Bars indicate standard deviation.

(ii) Lenga forests had exceptional regulating ES, and provided important provisioning ES for some industries (sawmills). Besides this, lenga forests had great values of supporting and cultural ES, and intermediate values of potential biodiversity compared to the other forest types. Similar to mixed evergreen forests, lenga were well represented inside the protection areas (69% of the forests were protected). Potential trade-offs can occur in these timber production forests outside the protection areas, especially for vulnerable species (e.g. *Hippocamelus bisulcus*) (Rosas et al., 2017).

(iii) Ñire forests had exceptional provisioning ES, due to silvopastoral methods, and intermediate values for some regulating and cultural ES. However, ñire forests had exceptional potential biodiversity compared to the other forest types, but was scarcely represented in the protection areas (only 14% of the forests were protected). In this sense, several potential trade-offs between ES and biodiversity exist because most of these forests occurred in private lands.

Martínez Pastur et al. (2017) found similar trends for ñire forests in Tierra del Fuego (Argentina), where the provision of the different ES greatly varied with the landscape: (i) provisioning ES were marginally greater in western than eastern, and northern than southern forests; (ii) regulating ES were higher in southern than northern forests; and (iii) cultural ES are influenced by the closeness to water bodies. Biodiversity was associated with site quality and the ecotone areas with lenga forests. As was described earlier, potential trade-offs were detected among the different ES (especially provisioning) and biodiversity.

3 Developing strategies of sustainable forest management

Temperate regions of the world have been extensively altered by human activities, with significant impacts on the provision of goods and services, and the loss of biodiversity (Franklin, 1988; Li et al., 2011; Lindenmayer et al., 2012). In southern Patagonia, sustainable management was mainly based on provisioning ES; however, during the recent years the concept of ES has allowed connections among economic, social, and ecological values to increase human well-being of local communities. In this context, new management strategies were proposed for ñire and lenga forests.

Nearly 70% of the ñire forests have been mainly used for livestock grazing, without clear conservation strategies or multipurpose management alternatives. This use lead to multiple trade-offs, that in many areas degraded the forest ecosystems and threatened forest continuity (i.e. over-browsing of regeneration). One alternative that combines multi-objectives, both economic and ecological, is SPS (Peri et al., 2016a). This method combines the three components, trees, grasslands or pastures, and livestock (sheep or cattle), in the same unit of land (Peri et al., 2016a). Forestry inventories, silvicultural practices, adjustment of stocking rates, and forest regeneration strategies have been developed in order to understand the dynamics among these three components, and maximize the benefits of forest management, and reduce the negative impacts of the interventions (Peri and Ormaechea, 2013). Livestock production is the main provisioning ES of the SPS (Peri et al., 2016a) linked to production of dry matter of grasslands and pastures that change through temperature, precipitation, nitrogen availability, and growth seasons (Peri et al., 2016b). Thinning was proposed as the main silvicultural treatment to increase timber wood and promote canopy opening for understory development. Besides this, SPS can maintain acceptable levels of (i) supporting ES compared to intensive management (e.g. clear-cuts in strips) or forest conversion to agriculture, habitat protection (e.g. bird refuge); (ii) regulation of ES such as carbon sequestration that ranged from 0.12 to 0.21 Mg-C ha^{-1}yr^{-1} (Peri et al., 2016a) or nutrient cycling (Bahamonde et al., 2012b); and (iii) to increase biodiversity conservation in the managed stands and protection of most of the characteristic species of the unmanaged forests (Peri et al., 2016b).

The traditional management in lenga forests was based on provisioning ES (mostly timber wood for sawmills), and mainly proposed to convert primary uneven-aged forests into secondary managed stands with a rotation length of 70–120 years depending on site quality and silviculture treatments (Gea Izquierdo et al., 2004). These proposals promoted the total removal of the original forest structure; clear-cuttings was prescribed in 40–50 m wide strips, as well as shelterwood cuts in two stages (first cut leaving 30 m^2 ha^{-1} basal area and a final cut that removes the remaining trees after 10–20 years) (Martínez Pastur et al., 2013). Nevertheless, these traditional practices significantly affected non-monetary ES and biodiversity (Soler et al., 2015, 2016). During recent years, new harvesting proposals have been needed for lenga forests to improve the land-sparing strategies at stand level due to local species extinctions (Deferrari et al., 2001; Spagarino et al., 2001; Martínez Pastur et al., 2002) and the social concerns about managing forests (Gamondés Moyano et al., 2016). The proposed method had yields comparable to traditional harvesting and improved the performance of cutting and skidding operations. Long-term plots quantified the conservation values and the variation in the natural cycles and ES provisions in the managed stands. Variable retention methods generated biotic and abiotic gradients in the managed stands, from closed canopies inside the aggregated retention to large openings in the dispersed retention outside the influence of the aggregates. These gradients allowed maintenance of biodiversity of the primary forests inside the aggregates and offered adequate conditions for other species including exotic and other species that live in the surrounding ecosystems. The natural cycles also followed the same pattern, where the same levels of primary forests were maintained in the aggregates but greatly changed in the harvested areas. Variable retention methods combine economic and ecological values, reaching to an equilibrium of both objectives in forest management and conservation planning (Martínez Pastur et al., 2009; Soler et al., 2015, 2016).

These management alternatives decreased the potential trade-offs among the provision of ES, and the other ES and biodiversity. The combination of land-sparing and land-sharing (Phalan et al., 2011) can help to achieve the multi-objective proposals for the forests achieving long-term sustainable management.

4 Forest ecosystem services in a changing world

The provision of ES must be resilient to variations in the long term, such as climate change (Lindenmayer and Franklin, 2002). The changes in climate are especially life-threatening in higher latitudes, where forests are more sensitive to small changes in climate that greatly influence the length of the growing season (Massaccesi et al., 2008; van Mantgem et al., 2009; Allen et al., 2010). Changes in climate conditions can alter the natural cycles of the forests and stand conditions by modifying the ES provisioning and conservation values of the forests. For example, supporting ES are directly linked to climate variations (e.g. net primary productivity) (Zhao and Running, 2010). In addition, these changes directly affect other ES by generating changes in secondary productivity or ecosystem functions and processes (Uhlenbrock Jansse and Rodríguez, 2005; Torres et al., 2015), such as regulating ES including water regulation, soil retention, and water purification.

Silviculture methods must be flexible and promote resilient managed forests (Gunderson, 2000; Elmqvist et al., 2003). Different harvesting intensities and retention strategies as well as landscape planning carried out to improve forest regeneration must at the same time maintain species assemblages, reduce the impact of exotic species, manage pests, and/or maintain timber production over time (Millar et al., 2007; Villalba et al., 2010; D'Amato et al., 2011; Torres et al., 2015).

Climate change brings new challenges for conservation strategies because it could modify the survival thresholds for many species. The design of nature reserves no longer provides for future survival of all the species for which they were designed (Swetnam et al., 1999; Heller and Zavaleta, 2009; Hallegatte, 2009; Gifford et al., 2011). Several studies proposed different alternatives to improve conservation strategies under climate change (Heller and Zavaleta, 2009), including the following: (i) design biologically complex reserves to capture unpredictable changes (Bartlein et al., 1997); (ii) increase the size of the reserves by covering greater climatic and environmental gradients (Hartig et al., 1997; Millar et al., 2007); (iii) increase the connectivity between natural areas (Halpin, 1997); (iv) conserve the species in more than one reserve (Halpin, 1997; Millar et al., 2007); (v) mitigate other impacts interacting with climate change impact (e.g., invasive species, fragmentation, contamination) (Chornesky et al., 2005); (vi) implement proposals based on adaptive management and long-term monitoring programs (Millar et al., 2007); and (vii) improve collaboration programs between different governmental institutions and stakeholders (e.g. NGOs) to adapt to changes at larger landscape scales (Grumbine, 1991).

5 Where to look for further information

For interest in ES and their implementation in sustainable proposals, we recommend searching for the results of projects that include multiple study sites, long-term monitoring, and models for implementation. For example, two large European Union projects defined the ES from different perspectives and explored methodologies to improve the ES characterization:

- OpenNESS (www.openness-project.eu), aimed to translate the concepts of natural capital and ES into operational frameworks with policy initiatives.
- OPERA (www.operas-project.eu), explored how and under what conditions these concepts can move beyond the academic domain toward practical implementation for sustainable ecosystem management.

Both projects developed one web site platform, OPPLA (oppla.eu), with open access designed for people with diverse needs and interests (from science, policy and practice, public, private and voluntary sectors, organizations large and small, as well as individuals). This platform provides knowledge of the marketplace, latest thinking on natural capital, ES, and nature-based solutions.

In particular, for those readers who are interested in temperate forests and southern Patagonian *Nothofagus* forests, we recommend the following:

- For broader and specific consideration about provision of ES and forest ecosystems (De Groot et al., 2002, 2010; Perera et al., 2018), or specific examples about ES provision (Martínez Pastur et al., 2016a; Peri et al., 2017, 2018, 2019).

- Survey of potential biodiversity at landscape level (Martínez Pastur et al., 2016b; Rosas et al., 2017, 2018) and invasive species (Anderson et al., 2011).
- Trade-offs between management and biodiversity conservation (Soler et al., 2015, 2016) and ES (Martínez Pasur et al., 2007, 2017; Peri et al., 2016b).
- Development of new forest management strategies considering ecological values (Martínez Pastur et al., 2009, 2013; Peri et al., 2016a) and climate change (Villalba et al., 2010; Kreps et al., 2012).

6 References

Allen, C. D., Macalady, A. K., Chenchouni, H., Bachelet, D., McDowell, N., Vennetier, M., Kitzberger, T., Rigling, A., Breshears, D. D., Hogg, E. H., Gonzalez, P., Fensham, R., Zhangm, Z., Castro, J., Demidova, N., Lim, J. H., Allard, G., Running, S. W., Semerci, A. and Cobb, N. 2010. A global overview of drought and heat-induced tree mortality reveals emerging climate change risks for forests. *For. Ecol. Manag.* 259(4), 660–84. doi:10.1016/j.foreco.2009.09.001.

Anderson, C. H., Soto, N., Cabello, J. L., Wallem, P., Martínez Pastur, G., Lencinas, M. V., Antúnez, D. and Davis, E. 2011. Building alliances between research and management to better control and mitigate the impacts of an invasive ecosystem engineer: the pioneering example of the North American beaver in the Fuegian Archipelago of Chile and Argentina. In: Francis, R. (Ed.), *A Handbook of Global Freshwater Invasive Species*. Earthscan Press, London, United Kingdom, pp. 347–59.

Andrew, M. E., Wulder, M. A. and Nelson, T. A. 2014. Potential contributions of remote sensing to ecosystem service assessments. *Prog. Phys. Geogr.* 38(3), 328–53. doi:10.1177/0309133314528942.

Armesto, J. J., Rozzi, R., Miranda, P. and Sabag, C. 1987. Plant/frugivore interactions in South American temperate forests. *Rev. Chil. Hist. Nat.* 60(2), 321–36.

Baillie, J. E. M., Hilton-Taylor, C. and Stuart, S. N. 2004. *IUCN Red List of Threatened Species: A Global Species Assessment*. IUCN, Gland, Switzerland.

Bartlein, P. J., Whitlock, C. and Shafter, S. L. 1997. Future climate in the Yellowstone National Park region and its potential impact on vegetation. *Conserv. Biol.* 11(3), 782–92. doi:10.1046/j.1523-1739.1997.95383.x.

Bennett, E. M., Peterson, G. D. and Gordon, L. J. 2009. Understanding relationships among multiple ecosystem services. *Ecol. Lett.* 12(12), 1394–404. doi:10.1111/j.1461-0248.2009.01387.x.

Bonino, N. 2005. *Guía de mamíferos de la Patagonia Argentina*. INTA Ed., Buenos Aires, Argentina.

Cardinale, B. J., Duffy, J. E., Gonzalez, A., Hooper, D. U., Perrings, C., Venail, P., Narwani, A., Mace, G. M., Tilman, D., Wardle, D. A., Kinzig, A. P., Daily, G. C., Loreau, M., Grace, J. B., Larigauderie, A., Srivastava, D. S. and Naeem, S. 2012. Biodiversity loss and its impact on humanity. *Nature* 486(7401), 59–67. doi:10.1038/nature11148.

Carvalho-Santos, C., Sousa-Silva, R., Goncalves, J. and Honrado, J. P. 2016. Ecosystem services and biodiversity conservation under forestation scenarios: options to improve management in the Vez watershed, NW Portugal. *Reg. Environ. Change* 16(6), 1557–70. doi:10.1007/s10113-015-0892-0.

Casalegno, S., Inger, R., De Silvey, C. and Gaston, K. J. 2013. Spatial covariance between aesthetic value and other ecosystem services. *PLoS One* 8(6), e68437. doi:10.1371/journal.pone.0068437.

Ceballos, G. and Ehrlich, P. R. 2002. Mammal population losses and the extinction crisis. *Science* 296(5569), 904–7. doi:10.1126/science.1069349.

Chen, H. Y. and Luo, Y. 2015. Net aboveground biomass declines of four major forest types with forest ageing and climate change in western Canada's boreal forests. *Glob. Chang. Biol.* 21(10), 3675–84. doi:10.1111/gcb.12994.

Chornesky, E. A., Bartuska, A. M., Aplet, G. H., Britton, K. O., Cummingscarlson, J., Davis, F. W., Eskow, J., Gordon, D. R., Gottschalk, K. W., Haack, R. A., Hansen, A. J., Mack, R. N., Rahel, F. J., Shannon, M. A., Wainger, L. A. and Wigley, T. B. 2005. Science priorities for reducing the threat of invasive species to sustainable forestry. *BioScience* 55(4), 335–48. doi:10.1641/0006-3568(2005)055[0335:SPFRTT]2.0.CO;2.

Cordingley, J. E., Newton, A. C., Rose, R. J., Clarke, R. T. and Bullock, J. M. 2016. Can landscape-scale approaches to conservation management resolve biodiversity-ecosystem service trade-offs? *J. Appl. Ecol.* 53(1), 96–105. doi:10.1111/1365-2664.12545.

D'Amato, A. W., Bradford, J. B., Fraver, S. and Palik, B. J. 2011. Forest management for mitigation and adaptation to climate change: insights from long-term silviculture experiments. *For. Ecol. Manag.* 262(5), 803–16. doi:10.1016/j.foreco.2011.05.014.

Daily, G. C. 1997. *Nature's Services.* Island Press, Washington DC.

Daily, G. C. and Ehrlich, P. R. 1995. Population diversity and the biodiversity crisis. In: Perrings, C., Maler, K., Folke, C., Holling, C. and Jansson, B. (Eds), *Biodiversity Conservation: Problems and Policies.* Kluwer Academic Press, Dordrecht, Holland, pp. 41–51.

Daniel, T. C., Muhar, A., Arnberger, A., Aznar, O., Boyd, J. W., Chan, K. M. A., Costanza, R., Elmqvist, T., Flint, C. G., Gobster, P. H., Grêt-Regamey, A., Lave, R., Muhar, S., Penker, M., Ribe, R. G., Schauppenlehner, T., Sikor, T., Soloviy, I., Spierenburg, M., Taczanowska, K., Tam, J. and von der Dunk, A. 2012. Cultural ecosystem services: potential contributions to the ecosystems services science and policy agenda. *Proc. Natl. Acad. Sci. U. S. A.* 109(23), 8812–9. doi:10.1073/pnas.1114773109.

de Bello, F., Lavorel, S., Díaz, S., Harrington, R., Cornelissen, J. H. C., Bardgett, R. D., Berg, M. P., Cipriotti, P., Feld, C. K., Hering, D., Martins da Silva, P., Potts, S. G., Sandin, L., Sousa, J. P., Storkey, J., Wardle, D. A. and Harrison, P. A. 2010. Towards an assessment of multiple ecosystem processes and services via functional traits. *Biodivers. Conserv.* 19(10), 2873–93. doi:10.1007/s10531-010-9850-9.

Deferrari, G., Camilion, C., Martínez Pastur, G. and Peri, P. L. 2001. Changes in *Nothofagus pumilio* forest biodiversity during the forest management cycle: birds. *Biodivers. Conserv.* 10(12), 2093–108. doi:10.1023/A:1013154824917.

De Groot, R. S., Alkemade, R., Braat, L., Hein, L. and Willemen, L. 2010. Challenges in integrating the concept of ecosystem services and values in landscape planning, management and decision making. *Ecol. Complexity* 7(3), 260–72. doi:10.1016/j.ecocom.2009.10.006.

De Groot, R. S., Wilson, M. A. and Boumans, R. M. J. 2002. A typology for the classification, description and valuation of ecosystem functions, goods and services. *Ecol. Econ.* 41(3), 393–408. doi:10.1016/S0921-8009(02)00089-7.

Díaz, S., Fargione, J., Chapin, F. S. and Tilman, D. 2006. Biodiversity loss threatens human well-being. *PLoS Biol.* 4(8), e277. doi:10.1371/journal.pbio.0040277.

Ducid, M. G., Murace, M. and Cellini, J. M. 2005. Diversidad fúngica en el filoplano de Osmorhiza spp. Relacionado con el sistema de regeneración empleado en bosques de *Nothofagus pumilio* en Tierra del Fuego, Argentina. *Bosque* 26(1), 33-42.

Durán, A. P., Duffy, J. P. and Gaston, K. J. 2014. Exclusion of agricultural lands in spatial conservation prioritization strategies: consequences for biodiversity and ecosystem service representation. *Proc. R. Soc. B* 281(1792), 1529. doi:10.1098/rspb.2014.1529.

Easterling, W. E., Brandle, J. R., Hays, C. J., Guo, Q. and Guertin, D. S. 2001. Simulating the impact of human land use change on forest composition in the Great Plains agroecosystems with the Seedscape model. *Ecol. Modell.* 140(1-2), 163-76. doi:10.1016/S0304-3800(01)00263-0.

Elith, J. and Leathwick, J. R. 2009. Species distribution models: ecological explanation and prediction across space and time. *Annu. Rev. Ecol. Evol. Syst.* 40(1), 677-97. doi:10.1146/annurev.ecolsys.110308.120159.

Elmqvist, T., Folke, C., Nystrom, M., Peterson, G., Bengtsson, J., Walker, B. and Norberg, J. 2003. Response diversity, ecosystem change, and resilience. *Front. Ecol. Environ.* 1(9), 488-94. doi:10.1890/1540-9295(2003)001[0488:RDECAR]2.0.CO;2.

Ernst, C., Gullick, R. and Nixon, K. 2004. Conserving forests to protect water. *Opflow* 30(5), 1-7. doi:10.1002/j.1551-8701.2004.tb01752.x.

Fischer, J., Batary, P., Bawa, K. S., Brussaard, L., Jahi Chappell, M., Clough, Y., Daily, G. C., Dorrough, J., Hartel, T., Jackson, L. E., Klein, A. M., Kremen, C., Kuemmerle, T., Lindenmayer, D. B., Mooney, H. A., Perfecto, I., Philpott, S. M., Tscharntke, T., Vandermeer, J., Cherico Wanger, T. and Von Wehrden, H. 2011. Conservation: limits of land sparing. *Science* 334(6056), 593.

Food and Agriculture Organization of the United Nations (FAO). 2010. Global forest resources assessment 2000. Technical Report. FAO, Rome, Italy.

Frangi, J. L., Richter, L. L., Barrera, M. D. and Alloggia, M. 1997. Decomposition of *Nothofagus* fallen woody debris in forests of Tierra del Fuego, Argentina. *Can. J. For. Res.* 27(7), 1095-102. doi:10.1139/x97-060.

Franklin, J. 1988. Structural and functional diversity in temperate forests. In: Wilson, E. O. and Peter, F. M. (Eds), *Biodiversity*. National Academies Press, Washington DC.

Gamondés Moyano, I. M., Morgan, R. K. and Martínez Pastur, G. 2016. Reshaping forest management in Southern Patagonia: a qualitative assessment. *J. Sustain. For.* 35(1), 37-59. doi:10.1080/10549811.2015.1043559.

Gargaglione, V., Peri, P. L. and Rubio, G. 2014. Tree-grass interactions for N in *Nothofagus antarctica* silvopastoral systems: evidence of facilitation from trees to underneath grasses. *Agrofor. Syst.* 88(5), 779-90. doi:10.1007/s10457-014-9724-3.

Gea Izquierdo, G., Martínez Pastur, G. M., Cellini, J. M. and Lencinas, M. V. 2004. Forty years of silvicultural management in southern *Nothofagus pumilio* primary forests. *For. Ecol. Manag.* 201(2-3), 335-47. doi:10.1016/j.foreco.2004.07.015.

Gifford, R., Kormos, C. and McIntyre, A. 2011. Behavioral dimensions of climate change: drivers, responses, barriers, and interventions. *Wiley Interdiscip. Rev. Clim. Change* 2(6), 801-27. doi:10.1002/wcc.143.

Goldman, R. L., Tallis, H., Kareiva, P. and Daily, G. C. 2008. Field evidence that ecosystem service projects support biodiversity and diversify options. *Proc. Natl. Acad. Sci. U. S. A.* 105(27), 9445-8. doi:10.1073/pnas.0800208105.

González, E., Salvo, A. and Valladares, G. 2015. Sharing enemies: evidence of forest contribution to natural enemy communities in crops, at different spatial scales. *Insect Conserv. Divers.* 8(4), 359–66. doi:10.1111/icad.12117.

Grove, S. and Meggs, J. 2003. Coarse woody debris, biodiversity and management: a review with particular reference to Tasmanian wet eucalypt forests. *Aust. For.* 66(4), 258–72. doi:10.1080/00049158.2003.10674920.

Grumbine, R. E. 1991. Cooperation or conflict-interagency relationships and the future of biodiversity for United-States parks and forests. *Environ. Manag.* 15(1), 27–37. doi:10.1007/BF02393836.

Guisan, A. and Zimmermann, N. E. 2000. Predictive habitat distribution models in ecology. *Ecol. Modell.* 135(2-3), 147–86. doi:10.1016/S0304-3800(00)00354-9.

Gunderson, L. H. 2000. Ecological resilience – in theory and application. *Annu. Rev. Ecol. Syst.* 31(1), 425–39. doi:10.1146/annurev.ecolsys.31.1.425.

Gustafsson, L., Baker, S. C., Bauhus, J., Beese, W. J., Brodie, A., Kouki, J., Lindenmayer, D. B., Lõhmus, A., Pastur, G. M., Messier, C., Neyland, M., Palik, B., Sverdrup-Thygeson, A., Volney, W. J. A., Wayne, A. and Franklin, J. F. 2012. Retention forestry to maintain multifunctional forests: a world perspective. *BioScience* 62(7), 633–45. doi:10.1525/bio.2012.62.7.6.

Haines-Young, R. and Potschin, M. 2010. The links between biodiversity, ecosystem services and human well-being. In: Raffaelli, D. G. and Frid, C. (Eds), *Ecosystem Ecology. A New Synthesis.* Cambridge University Press, New York, pp. 110–39.

Hallegatte, S. 2009. Strategies to adapt to an uncertain climate change. *Glob. Environ. Chang.* 19(2), 240–7. doi:10.1016/j.gloenvcha.2008.12.003.

Halpin, P. N. 1997. Global climate change and natural-area protection: management responses and research directions. *Ecol. Appl.* 7(3), 828–43. doi:10.1890/1051-07 61(1997)007[0828:GCCANA]2.0.CO;2.

Harcourt, A. H., Parks, S. A. and Woodroffe, R. 2001. Human density as an influence on species/area relationships: double jeopardy for small African reserves? *Biodivers. Conserv.* 10(6), 1011–26. doi:10.1023/A:1016680327755.

Harrison, P. A., Berry, P. M., Simpson, G., Haslett, J. R., Blicharska, M., Bucur, M., Dunford, R., Egoh, B., Garcia-Llorente, M., Geamănă, N., Geertsema, W., Lommelen, E., Meiresonne, L. and Turkelboom, F. 2014. Linkages between biodiversity attributes and ecosystem services: a systematic review. *Ecosyst. Serv.* 9, 191–203. doi:10.1016/j. ecoser.2014.05.006.

Hartig, E. K., Grozev, O. and Rosenzweig, C. 1997. Climate change, agriculture and wetlands in Eastern Europe: vulnerability, adaptation and policy. *Clim. Chang.* 36(1-2), 107–21.

Heller, N. E. and Zavaleta, E. S. 2009. Biodiversity management in the face of climate change: a review of 22 years of recommendations. *Biol. Conserv.* 142(1), 14–32. doi:10.1016/j.biocon.2008.10.006.

Henríquez, J. M. and Lusk, C. H. 2005. Facilitation of *Nothofagus antarctica* (Fagaceae) seedlings by the prostrate shrub *Empetrum rubrum* (Empetraceae) on glacial moraines in Patagonia. *Austral. Ecol.* 30(8), 877–82. doi:10.1111/j. 1442-9993.2005.01531.x.

Hewitt, R. E., Taylor, D. L., Hollingsworth, T. N., Anderson, C. B. and Martínez Pastur, G. 2018. Variable retention harvesting influences belowground plant-fungal interactions of *Nothofagus pumilio* seedlings in forests of southern Patagonia. *Peer J.* 6, e5008. doi:10.7717/peerj.5008.

Hirzel, A. H., Hausser, J., Chessel, D. and Perrin, N. 2002. Ecological-niche factor analysis: how to compute habitat- suitability maps without absence data? *Ecology* 83(7), 2027–36. doi:10.1890/0012-9658(2002)083[2027:ENFAHT]2.0.CO;2.

Hudson, L. N., Newbold, T., Contu, S., Hill, S. L., Lysenko, I., De Palma, A., Phillips, H. R., Alhusseini, T. I., Bedford, F. E., Bennett, D. J., Booth, H., Burton, V. J., Chng, C. W., Choimes, A., Correia, D. L., Day, J., Echeverría-Londoño, S., Emerson, S. R., Gao, D., Garon, M., Harrison, M. L., Ingram, D. J., Jung, M., Kemp, V., Kirkpatrick, L., Martin, C. D., Pan, Y., Pask-Hale, G. D., Pynegar, E. L., Robinson, A. N., Sanchez-Ortiz, K., Senior, R. A., Simmons, B. I., White, H. J., Zhang, H., Aben, J., Abrahamczyk, S., Adum, G. B., Aguilar-Barquero, V., Aizen, M. A., Albertos, B., Alcala, E. L., Del Mar Alguacil, M., Alignier, A., Ancrenaz, M., Andersen, A. N., Arbeláez-Cortés, E., Armbrecht, I., Arroyo-Rodríguez, V., Aumann, T., Axmacher, J. C., Azhar, B., Azpiroz, A. B., Baeten, L., Bakayoko, A., Báldi, A., Banks, J. E., Baral, S. K., Barlow, J., Barratt, B. I., Barrico, L., Bartolommei, P., Barton, D. M., Basset, Y., Batáry, P., Bates, A. J., Baur, B., Bayne, E. M., Beja, P., Benedick, S., Berg, Å, Bernard, H., Berry, N. J., Bhatt, D., Bicknell, J. E., Bihn, J. H., Blake, R. J., Bobo, K. S., Bóçon, R., Boekhout, T., Böhning-Gaese, K., Bonham, K. J., Borges, P. A., Borges, S. H., Boutin, C., Bouyer, J., Bragagnolo, C., Brandt, J. S., Brearley, F. Q., Brito, I., Bros, V., Brunet, J., Buczkowski, G., Buddle, C. M., Bugter, R., Buscardo, E., Buse, J., Cabra-García, J., Cáceres, N. C., Cagle, N. L., Calviño-Cancela, M., Cameron, S. A., Cancello, E. M., Caparrós, R., Cardoso, P., Carpenter, D., Carrijo, T. F., Carvalho, A. L., Cassano, C. R., Castro, H., Castro-Luna, A. A., Rolando, C. B., Cerezo, A., Chapman, K. A., Chauvat, M., Christensen, M., Clarke, F. M., Cleary, D. F., Colombo, G., Connop, S. P., Craig, M. D., Cruz-López, L., Cunningham, S. A., D'Aniello, B., D'Cruze, N., da Silva, P. G., Dallimer, M., Danquah, E., Darvill, B., Dauber, J., Davis, A. L., Dawson, J., de Sassi, C., de Thoisy, B., Deheuvels, O., Dejean, A., Devineau, J. L., Diekötter, T., Dolia, J. V., Domínguez, E., Dominguez-Haydar, Y., Dorn, S., Draper, I., Dreber, N., Dumont, B., Dures, S. G., Dynesius, M., Edenius, L., Eggleton, P., Eigenbrod, F., Elek, Z., Entling, M. H., Esler, K. J., de Lima, R. F., Faruk, A., Farwig, N., Fayle, T. M., Felicioli, A., Felton, A. M., Fensham, R. J., Fernandez, I. C., Ferreira, C. C., Ficetola, G. F., Fiera, C., Filgueiras, B. K., Fırıncıoğlu, H. K., Flaspohler, D., Floren, A., Fonte, S. J., Fournier, A., Fowler, R. E., Franzén, M., Fraser, L. H., Fredriksson, G. M., Freire, G. B., Frizzo, T. L., Fukuda, D., Furlani, D., Gaigher, R., Ganzhorn, J. U., García, K. P., Garcia-R, J. C., Garden, J. G., Garilleti, R., Ge, B. M., Gendreau-Berthiaume, B., Gerard, P. J., Gheler-Costa, C., Gilbert, B., Giordani, P., Giordano, S., Golodets, C., Gomes, L. G., Gould, R. K., Goulson, D., Gove, A. D., Granjon, L., Grass, I., Gray, C. L., Grogan, J., Gu, W., Guardiola, M., Gunawardene, N. R., Gutierrez, A. G., Gutiérrez-Lamus, D. L., Haarmeyer, D. H., Hanley, M. E., Hanson, T., Hashim, N. R., Hassan, S. N., Hatfield, R. G., Hawes, J. E., Hayward, M. W., Hébert, C., Helden, A. J., Henden, J. A., Henschel, P., Hernández, L., Herrera, J. P., Herrmann, F., Herzog, F., Higuera-Diaz, D., Hilje, B., Höfer, H., Hoffmann, A., Horgan, F. G., Hornung, E., Horváth, R., Hylander, K., Isaacs-Cubides, P., Ishida, H., Ishitani, M., Jacobs, C. T., Jaramillo, V. J., Jauker, B., Hernández, F. J., Johnson, M. F., Jolli, V., Jonsell, M., Juliani, S. N., Jung, T. S., Kapoor, V., Kappes, H., Kati, V., Katovai, E., Kellner, K., Kessler, M., Kirby, K. R., Kittle, A. M., Knight, M. E., Knop, E., Kohler, F., Koivula, M., Kolb, A., Kone, M., Kőrösi, Á, Krauss, J., Kumar, A., Kumar, R., Kurz, D. J., Kutt, A. S., Lachat, T., Lantschner, V., Lara, F., Lasky, J. R., Latta, S. C., Laurance, W. F., Lavelle, P., Le Féon, V., LeBuhn, G., Légaré, J. P., Lehouck, V., Lencinas, M. V., Lentini, P. E., Letcher, S. G., Li, Q., Litchwark, S. A., Littlewood, N. A., Liu, Y., Lo-Man-Hung, N., López-Quintero, C. A., Louhaichi, M.,

Lövei, G. L., Lucas-Borja, M. E., Luja, V. H., Luskin, M. S., MacSwiney G, M. C., Maeto, K., Magura, T., Mallari, N. A., Malone, L. A., Malonza, P. K., Malumbres-Olarte, J., Mandujano, S., Måren, I. E., Marin-Spiotta, E., Marsh, C. J., Marshall, E. J., Martínez, E., Martínez Pastur, G., Moreno Mateos, D., Mayfield, M. M., Mazimpaka, V., McCarthy, J. L., McCarthy, K. P., McFrederick, Q. S., McNamara, S., Medina, N. G., Medina, R., Mena, J. L., Mico, E., Mikusinski, G., Milder, J. C., Miller, J. R., Miranda-Esquivel, D. R., Moir, M. L., Morales, C. L., Muchane, M. N., Muchane, M., Mudri-Stojnic, S., Munira, A. N., Muoñz-Alonso, A., Munyekenye, B. F., Naidoo, R., Naithani, A., Nakagawa, M., Nakamura, A., Nakashima, Y., Naoe, S., Nates-Parra, G., Navarrete Gutierrez, D. A., Navarro-Iriarte, L., Ndang'ang'a, P. K., Neuschulz, E. L., Ngai, J. T., Nicolas, V., Nilsson, S. G., Noreika, N., Norfolk, O., Noriega, J. A., Norton, D. A., Nöske, N. M., Nowakowski, A. J., Numa, C., O'Dea, N., O'Farrell, P. J., Oduro, W., Oertli, S., Ofori-Boateng, C., Oke, C. O., Oostra, V., Osgathorpe, L. M., Otavo, S. E., Page, N. V., Paritsis, J., Parra-H, A., Parry, L., Pe'er, G., Pearman, P. B., Pelegrin, N., Pélissier, R., Peres, C. A., Peri, P. L., Persson, A. S., Petanidou, T., Peters, M. K., Pethiyagoda, R. S., Phalan, B., Philips, T. K., Pillsbury, F. C., Pincheira-Ulbrich, J., Pineda, E., Pino, J., Pizarro-Araya, J., Plumptre, A. J., Poggio, S. L., Politi, N., Pons, P., Poveda, K., Power, E. F., Presley, S. J., Proença, V., Quaranta, M., Quintero, C., Rader, R., Ramesh, B. R., Ramirez-Pinilla, M. P., Ranganathan, J., Rasmussen, C., Redpath-Downing, N. A., Reid, J. L., Reis, Y. T., Rey Benayas, J. M., Rey-Velasco, J. C., Reynolds, C., Ribeiro, D. B., Richards, M. H., Richardson, B. A., Richardson, M. J., Ríos, R. M., Robinson, R., Robles, C. A., Römbke, J., Romero-Duque, L. P., Rös, M., Rosselli, L., Rossiter, S. J., Roth, D. S., Roulston, T. H., Rousseau, L., Rubio, A. V., Ruel, J. C., Sadler, J. P., Sáfián, S., Saldaña-Vázquez, R. A., Sam, K., Samnegård, U., Santana, J., Santos, X., Savage, J., Schellhorn, N. A., Schilthuizen, M., Schmiedel, U., Schmitt, C. B., Schon, N. L., Schüepp, C., Schumann, K., Schweiger, O., Scott, D. M., Scott, K. A., Sedlock, J. L., Seefeldt, S. S., Shahabuddin, G., Shannon, G., Sheil, D., Sheldon, F. H., Shochat, E., Siebert, S. J., Silva, F. A., Simonetti, J. A., Slade, E. M., Smith, J., Smith-Pardo, A. H., Sodhi, N. S., Somarriba, E. J., Sosa, R. A., Soto Quiroga, G., St-Laurent, M. H., Starzomski, B. M., Stefanescu, C., Steffan-Dewenter, I., Stouffer, P. C., Stout, J. C., Strauch, A. M., Struebig, M. J., Su, Z., Suarez-Rubio, M., Sugiura, S., Summerville, K. S., Sung, Y. H., Sutrisno, H., Svenning, J. C., Teder, T., Threlfall, C. G., Tiitsaar, A., Todd, J. H., Tonietto, R. K., Torre, I., Tóthmérész, B., Tscharntke, T., Turner, E. C., Tylianakis, J. M., Uehara-Prado, M., Urbina-Cardona, N., Vallan, D., Vanbergen, A. J., Vasconcelos, H. L., Vassilev, K., Verboven, H. A., Verdasca, M. J., Verdú, J. R., Vergara, C. H., Vergara, P. M., Verhulst, J., Virgilio, M., Vu, L. V., Waite, E. M., Walker, T. R., Wang, H. F., Wang, Y., Watling, J. I., Weller, B., Wells, K., Westphal, C., Wiafe, E. D., Williams, C. D., Willig, M. R., Woinarski, J. C., Wolf, J. H., Wolters, V., Woodcock, B. A., Wu, J., Wunderle, J. M., Yamaura, Y., Yoshikura, S., Yu, D. W., Zaitsev, A. S., Zeidler, J., Zou, F., Collen, B., Ewers, R. M., Mace, G. M., Purves, D. W., Scharlemann, J. P. and Purvis, A. 2017. The database of the PREDICTS (Projecting Responses of Ecological Diversity in Changing Terrestrial Systems) project. *Ecol. Evol.* 7(1), 145–88. doi:10.1002/ece3.2579.

Ippi, S., Anderson, C. B., Rozzi, R. and Elphick, C. 2009. Annual variation of abundance and composition in forest bird assemblages on Navarino Island, Cape Horn Biosphere Reserve, Chile. *Ornitol. Neotrop.* 20(2), 231–45.

Kreps, G., Martínez Pastur, G. and Peri, P. L. 2012. *Cambio climático en Patagonia Sur: escenarios futuros en el manejo de los recursos naturales.* INTA Ed., Buenos Aires, Argentina.

Kreye, M., Adams, D. and Escobedo, F. 2014. The value of forest conservation for water quality protection. *Forests* 5(5), 862–84. doi:10.3390/f5050862.

Lal, R. and Lorenz, K. 2012. Carbon sequestration in temperate forests. In: Lal, R., Lorenz, K., Hüttl, R., Schneider, B. and von Braun, J. (Eds), *Recarbonization of the Biosphere*. Springer, Amsterdam, Holland, pp. 187–202.

Lanfranco, D. 1977. Entomofauna asociada a los bosques de *Nothofagus pumilio* en la región de Magallanes 1parte: Monte Alto (Río Rubens, Última Esperanza). *Ann. Inst. Pat.* 8(1), 319–46.

Lencinas, M. V., Martínez Pastur, G. M., Medina, M. and Busso, C. 2005. Richness and density of birds in timber *Nothofagus pumilio* forests and their unproductive associated environments. *Biodivers. Conserv.* 14(10), 2299–320. doi:10.1007/s10531-004-1665-0.

Lencinas, M. V., Martínez Pastur, G., Gallo, E., Moretto, A., Busso, C. and Peri, P. L. 2007. Mitigation of biodiversity loss in *Nothofagus pumilio* managed forests of South Patagonia. In: Pacha, M. J., Luque, S., Galetto, L. and Iverson, L. (Eds), *Understanding Biodiversity Loss: an Overview of Forest Fragmentation in South America, Part III: Landscape Ecology for Conservation, Management and Restoration*. IALE Landscape Research and Management Papers, Grenoble, France, pp. 112–20.

Lencinas, M. V., Martínez Pastur, G., Rivero, P. and Busso, C. 2008a. Conservation value of timber quality vs. associated non-timber quality stands for understory diversity in *Nothofagus* forests. *Biodivers. Conserv.* 17(11), 2579–97. doi:10.1007/s10531-008-9323-6.

Lencinas, M. V., Martínez Pastur, G. M., Anderson, C. B. and Busso, C. 2008b. The value of timber quality forests for insect conservation on Tierra del Fuego Island compared to associated non-timber quality stands. *J. Insect Conserv.* 12(5), 461–75. doi:10.1007/s10841-007-9079-4.

Lencinas, M. V., Martínez Pastur, G., Solán, R., Gallo, E. and Cellini, J. M. 2008c. Forest management with variable retention impact over moss communities of *Nothofagus pumilio* understory. *Forstarchiv* 79(1), 77–82.

Lencinas, M. V., Martínez Pastur, G. M., Gallo, E. and Cellini, J. M. 2009. Alternative silvicultural practices with variable retention improve bird conservation in managed South Patagonian forests. *For. Ecol. Manag.* 258(4), 472–80. doi:10.1016/j.foreco.2009.01.012.

Li, C., Raffaele, L. and Chen, J. 2011. *Landscape Ecology in Forest Management and Conservation*. Springer, Berlin, Heidelberg.

Lindenmayer, D. and Franklin, J. 2002. *Conserving Forest Biodiversity: a Comprehensive Multiscaled Approach*. Island Press, Washington DC.

Lindenmayer, D. B., Franklin, J. F., Lõhmus, A., Baker, S. C., Bauhus, J., Beese, W., Brodie, A., Kiehl, B., Kouki, J., Martínez Pastur, G. M., Messier, C., Neyland, M., Palik, B., Sverdrup-Thygeson, A., Volney, J., Wayne, A. and Gustafsson, L. 2012. A major shift to the retention approach for forestry can help resolve some global forest sustainability issues. *Conserv. Lett.* 5(6), 421–31. doi:10.1111/j.1755-263X.2012.00257.x.

Loreau, M. and de Mazancourt, C. 2013. Biodiversity and ecosystem stability: a synthesis of underlying mechanisms. *Ecol. Lett.* 16(1), 106–15. doi:10.1111/ele.12073.

Mace, G. M., Norris, K. and Fitter, A. H. 2012. Biodiversity and ecosystem services: a multilayered relationship. *Trends Ecol. Evol. (Amst.)* 27(1), 19–26. doi:10.1016/j.tree.2011.08.006.

Maes, J., Teller, A., Erhard, M., Murphy, Paracchini, M. L., Barredo, J. I., Grizzetti, B., Cardoso, A., Somma, F., Petersen, J. E., Meiner, A., Royo Gelabert, E., Zal, N., Kristensen, P., Bastrup-Birk, A., Biala, K., Romao, C., Piroddi, C., Egoh, B. C., Florina, Santos-Martín, F., Naruševičius, V., Verboven, J., Pereira, H. M., Bengtsson, J., Gocheva, K., Marta-Pedroso, C., Snäll, T., Estreguil, C., San-Miguel-Ayanz, J., Braat, L., Grêt-Regamey, A., Pérez-Soba, M., Degeorges, P., Beaufaron, C., Lillebø, A. I., Abdul Malak, D., Liquete, C., Condé, S., Moen, J., Ostergard, H., Czúcz, B., Drakou, E. G., Zulian, G. and Lavalle, C.. 2014. Mapping and Assessment of Ecosystems and their Services: Indicators for ecosystem assessments under Action 5 of the EU Biodiversity Strategy to 2020. European Union Technical Report No. 2014-080. Luxembourg.

Martínez Pastur, G., Peri, P. L., Fernández, C., Staffieri, G. and Lencinas, M. V. 2002. Changes in understory species diversity during the *Nothofagus pumilio* forest management cycle. *J. For. Res.* 7(3), 165–74.

Martínez Pastur, G., Lencinas, M. V., Peri, P. L., Moretto, A., Cellini, J. M., Mormeneo, I. and Vukasovic, R. 2007. Harvesting adaptation to biodiversity conservation in sawmill industry: technology innovation and monitoring program. *International Journal of Innovation Management and Technology* 2(3), 58–70.

Martínez Pastur, G., Lencinas, M. V., Cellini, J. M., Peri, P. L. and Soler Esteban, R. 2009. Timber management with variable retention in *Nothofagus pumilio* forests of Southern Patagonia. *For. Ecol. Manag.* 258(4), 436–43. doi:10.1016/j.foreco.2009.01.048.

Martínez Pastur, G., Peri, P. L., Lencinas, M. V., Cellini, J. M., Barrera, M., Soler Esteban, R., Ivancich, H., Mestre, L., Moretto, A., Anderson, C. and Pulido, F. 2013. La producción forestal y la conservación de la biodiversidad en los bosques de *Nothofagus* en Tierra del Fuego y Patagonia Sur. In: Donoso, P. and Promis, A. (Eds), *Silvicultura en bosques nativos: Avances en la investigación en Chile, Argentina y Nueva Zelanda*. Universidad Austral de Chile, Valdivia, Chile, pp. 155–79.

Martínez Pastur, G. M., Lencinas, M. V., Gallo, E., De Cruz, M., Borla, M. L., Esteban, R. S. and Anderson, C. B. 2015. Habitat-specific vegetation and seasonal drivers of bird community structure and function in southern Patagonian forests. *Commun. Ecol.* 16(1), 55–65. doi:10.1556/168.2015.16.1.7.

Martínez Pastur, G., Peri, P. L., Lencinas, M. V., García-Llorente, M. and Martín-López, B. 2016a. Spatial patterns of cultural ecosystem services provision in Southern Patagonia. *Landsc. Ecol.* 31(2), 383–99. doi:10.1007/s10980-015-0254-9.

Martínez Pastur, G., Peri, P. L., Soler, R. M., Schindler, S. and Lencinas, M. V. 2016b. Biodiversity potential of *Nothofagus* forests in Tierra del Fuego (Argentina): tool proposal for regional conservation planning. *Biodivers. Conserv.* 25(10), 1843–62. doi:10.1007/s10531-016-1162-2.

Martínez Pastur, G., Peri, P. L., Huertas Herrera, A., Schindler, S., Díaz-Delgado, R., Lencinas, M. V. and Soler, R. 2017. Linking potential biodiversity and three ecosystem services in silvopastoral managed forest landscapes of Tierra del Fuego, Argentina. *International Journal of Biodiversity Science, Ecosystem Services and Management* 13(2), 1–11. doi:10.1080/21513732.2016.1260056.

Martins, K. T., Gonzalez, A. and Lechowicz, M. J. 2015. Pollination services are mediated by bee functional diversity and landscape context. *Agric. Ecosyst. Environ.* 200, 12–20. doi:10.1016/j.agee.2014.10.018.

Maskell, L. C., Crowe, A., Dunbar, M. J., Emmett, B., Henrys, P., Keith, A. M., Norton, L. R., Scholefield, P., Clark, D. B., Simpson, I. C. and Smart, S. M. 2013. Exploring the ecological constraints to multiple ecosystem service delivery and biodiversity. *J. Appl. Ecol.* 50(3), 561–71. doi:10.1111/1365-2664.12085.

Massaccesi, G., Roig, F. A., Martínez Pastur, G. J. and Barrera, M. D. 2008. Growth patterns of *Nothofagus pumilio* trees along altitudinal gradients in Tierra del Fuego, Argentina. *Trees* 22(2), 245–55. doi:10.1007/s00468-007-0181-8.

McQuillan, P. B. 1993. *Nothofagus* (Fagaceae) and its invertebrate fauna – an overview and preliminary synthesis. *Biol. J. Linn. Soc.* 49(4), 317–54. doi:10.1111/j.1095-8312.1993. tb00910.x.

Millar, C. I., Stephenson, N. L. and Stephens, S. L. 2007. Climate change and forests of the future: managing in the face of uncertainty. *Ecol. Appl.* 17(8), 2145–51. doi:10.1890/06-1715.1.

Millennium Ecosystem Assessment (MEA). 2005. *Ecosystems and Human Wellbeing: Current State and Trends*. Island Press, Washington DC.

Morales, C. L. and Aizén, M. A. 2006. Invasive mutualisms and the structure of plant-pollinator interactions in the temperate forests of north-west Patagonia, Argentina. *J. Ecol.* 94(1), 171–80. doi:10.1111/j.1365-2745.2005.01069.x.

Mori, A. S., Lertzman, K. P. and Gustafsson, L. 2017. Biodiversity and ecosystem services in forest ecosystems: a research agenda for applied forest ecology. *J. Appl. Ecol.* 54(1), 12–27. doi:10.1111/1365-2664.12669.

Müller, J., Boch, S., Blaser, S., Fischer, M. and Prati, D. 2015. Effects of forest management on bryophyte communities on deadwood. *Nova Hedwigia* 100(3), 423–38. doi:10.1127/nova_hedwigia/2015/0242.

Panagos, P., Borrelli, P., Poesen, J., Ballabio, C., Lugato, E., Meusburger, K., Montanarella, L. and Alewell, C. 2015. The new assessment of soil loss by water erosion in Europe. *Environ. Sci. Policy* 54, 438–47. doi:10.1016/j.envsci.2015.08.012.

Perera, A., Peterson, U., Martínez Pastur, G. and Iverson, L. 2018. *Ecosystem Services from Forest Landscapes: Broadscale Considerations*. Ed. Springer, Cham, Switzerland.

Peri, P. L. and Ormaechea, S. 2013. *Relevamiento de los bosques nativos de ñire (Nothofagus antarctica) en Santa Cruz: base para su conservación y manejo*. INTA Ed., Río Gallegos, Argentina.

Peri, P. L., Gargaglione, V., Martínez Pastur, G. and Lencinas, M. V. 2010. Carbon accumulation along a stand development sequence of *Nothofagus antarctica* forests across a gradient in site quality in Southern Patagonia. *For. Ecol. Manag.* 260(2), 229–37. doi:10.1016/j.foreco.2010.04.027.

Peri, P. L., Hansen, N. E., Bahamonde, H. A., Lencinas, M. V., Von Müller, A. R., Ormaechea, S., Gargaglione, V., Soler Esteban, R., Tejera, L. E., Lloyd, C. E. and Martínez Pastur, G. M. 2016a. Silvopastoral systems under native forest in Patagonia, Argentina. In: Peri, P. L., Dube, F. and Varella, A. (Eds), *Silvopastoral Systems in Southern South America*. Springer. Series: Advances in Agroforestry. Springer, Bern, Switzerland, pp. 117–68.

Peri, P. L., Ladd, B., Lasagno, R. G. and Martínez Pastur, G. 2016b. The effects of land management (grazing intensity) vs. the effects of topography, soil properties, vegetation type, and climate on soil carbon concentration in Southern Patagonia. *J. Arid Environ.* 134, 73–8. doi:10.1016/j.jaridenv.2016.06.017.

Peri, P. L., Banegas, N., Gasparri, I., Carranza, C., Rossner, B., Martínez Pastur, G., Caballero, L., López, D., Loto, D., Fernández, P., Powel, P., Ledesma, M., Pedraza, R., Albanesi, A., Bahamonde, H. A., Eclesia, R. and Piñeiro, G. 2017. Carbon sequestration in temperate silvopastoral systems, Argentina. In: Montagnini, F. (Ed.), *Integrating*

Landscapes: Agroforestry for Biodiversity Conservation and Food Sovereignty.
Springer Series: Advances in Agroforestry. Springer, Bern, Suiza, pp. 453–78.

Peri, P., Rosas, Y. M., Ladd, B., Toledo, S., Lasagno, R. and Martínez Pastur, G. 2018.
Modelling soil carbon content in South Patagonia and evaluating changes
according to climate, vegetation, desertification and grazing. *Sustainability* 10(2),
438. doi:10.3390/su10020438.

Peri, P. L., Lasagno, R. G., Martínez Pastur, G., Atkinson, R., Thomas, E. and Ladd, B. 2019.
Soil carbon is a useful surrogate for conservation planning in developing nations.
Sci. Rep. 9(1), 3905. doi:10.1038/s41598-019-40741-0.

Phalan, B., Onial, M., Balmford, A. and Green, R. E. 2011. Reconciling food production
and biodiversity conservation: land sharing and land sparing compared. *Science*
333(6047), 1289–91. doi:10.1126/science.1208742.

Premoli, A. C. and Brewer, C. A. 2007. Environmental v. genetically driven variation in
ecophysiological traits of *Nothofagus pumilio* from contrasting elevations. *Aust. J.
Bot.* 55(6), 585–91. doi:10.1071/BT06026.

Premoli, A. C., Raffaele, E. and Mathiasen, P. 2007. Morphological and phenological
differences in *Nothofagus pumilio* from contrasting elevations: evidence from a
common garden. *Austral. Ecol.* 32(5), 515–23. doi:10.1111/j.1442-9993.2007.01720.x.

Prescott, C. E. 2002. The influence of the forest canopy on nutrient cycling. *Tree Physiol.*
22(15–16), 1193–200. doi:10.1093/treephys/22.15-16.1193.

Quintas-Soriano, C., Martín-López, B., Santos-Martín, F., Loureiro, M., Montes, C., Benayas,
J. and García-Llorente, M. 2016. Ecosystem services values in Spain: a meta-analysis.
Environ. Sci. Policy 55(01), 186–95. doi:10.1016/j.envsci.2015.10.001.

Rabassa, J., Coronato, A., Bujalesky, G., Salemme, M., Roig, C., Meglioli, A., Heusser, C.,
Gordillo, S., Roig, F., Borromei, A. and Quattrocchio, M. 2000. Quaternary of Tierra
del Fuego, Southernmost South America: an updated review. *Quart. Int.* 68(71),
217–40.

Ramírez, C., San Martín, C., Oyarzún, A. and Figueroa, H. 1997. Morpho-ecological study
on the South American species of the genus *Nothofagus. Plant Ecol.* 130(2), 101–9.
doi:10.1023/A:1009735821549.

Rands, M. R., Adams, W. M., Bennun, L., Butchart, S. H., Clements, A., Coomes, D., Entwistle,
A., Hodge, I., Kapos, V., Scharlemann, J. P. W., Sutherland, W. J. and Vira, B. 2010.
Biodiversity conservation: challenges beyond 2010. *Science* 329(5997), 1298–303.
doi:10.1126/science.1189138.

Raudsepp-Hearne, C., Peterson, G. D. and Bennett, E. M. 2010. Ecosystem service bundles
for analyzing tradeoffs in diverse landscapes. *Proc. Natl. Acad. Sci. U.S.A.* 107(11),
5242–7. doi:10.1073/pnas.0907284107.

Reyers, B., Biggs, R., Cumming, G. S., Elmqvist, T., Hejnowicz, A. P. and Polasky, S. 2013.
Getting the measure of ecosystem services: a social-ecological approach. *Front.
Ecol. Environ.* 11(5), 268–73. doi:10.1890/120144.

Rodríguez, J. P., Brotons, L., Bustamante, J. and Seoane, J. 2007. The application of
predictive modelling of species distribution to biodiversity conservation. *Divers.
Distrib.* 13(3), 243–51. doi:10.1111/j.1472-4642.2007.00356.x.

Rosas, Y. M., Peri, P. L., Huertas Herrera, A., Pastore, H. and Martínez Pastur, G. 2017.
Modeling of potential habitat suitability of *Hippocamelus bisulcus*: effectiveness of a
protected areas network in Southern Patagonia. *Ecol. Proc.* 6(1), 28–42. doi:10.1186/
s13717-017-0096-2.

Rosas, Y. M., Peri, P. L. and Martínez Pastur, G. 2018. Potential biodiversity map of lizard species in Southern Patagonia: environmental characterization, desertification influence and analyses of protection areas. *Amphib.-Reptilia* 39(3), 289–301. doi:10.1163/15685381-20181001.

Russell, R., Guerry, A. D., Balvanera, P., Gould, R. K., Basurto, X., Chan, K. M. A., Klain, S., Levine, J. and Tam, J. 2013. Humans and nature: how knowing and experiencing nature affect well-being. *Annu. Rev. Environ. Resour.* 38(1), 473–502. doi:10.1146/annurev-environ-012312-110838.

Sanderson, E. W., Jaiteh, M., Levy, M. A., Redford, K. H., Wannebo, A. V. and Woolmer, G. 2002. The human footprint and the last of the wild: the human footprint is a global map of human influence on the land surface, which suggests that human beings are stewards of nature, whether we like it or not. *BioScience* (52)(10), 891–904.

Schindler, S., Sebesvari, Z., Damm, C., Euller, K., Mauerhofer, V., Schneidergruber, A., Biró, M., Essl, F., Kanka, R., Lauwaars, S. G., Schulz-Zunkel, C., van der Sluis, T., Kropik, M., Gasso, V., Krug, A., T. Pusch, M., Zulka, K. P., Lazowski, W., Hainz-Renetzeder, C., Henle, K. and Wrbka, T. 2014. Multifunctionality of floodplain landscapes: relating management options to ecosystem services. *Landsc. Ecol.* 29(2), 229–44. doi:10.1007/s10980-014-9989-y.

Schindler, S., O'Neill, F. H., Biró, M., Damm, C., Gasso, V., Kanka, R., van der Sluis, T., Krug, A., Lauwaars, S. G., Sebesvari, Z., Pusch, M., Baranocsky, B., Ehlert, T., Neukirchen, B., Martin, J. R., Euller, K., Mauerhofer, V. and Webka, T. 2016. Multifunctional floodplain management and biodiversity effects: a knowledge synthesis for six European countries. *Biodivers. Conserv.* 25(7), 1349–82. doi:10.1007/s10531-016-1129-3.

Soberón, J. and Peterson, A. T. 2005. Interpretation of models of fundamental ecological niches and species' distribution areas. *Biodiv. Inform.* 2(1), 1–10.

Soler, R., Martínez Pastur, G., Lencinas, M. V. and Borrelli, L. 2012. Differential forage use between native and domestic herbivores in southern Patagonian *Nothofagus* forests. *Agrofor. Syst.* 85(3), 397–409.

Soler, R., Schindler, S., Lencinas, M. V., Peri, P. L. and Martínez Pastur, G. 2015. Retention forestry in southern Patagonia: multiple environmental impacts and their temporal trends. *Int. For. Rev.* 17(2), 231–43.

Soler, R. M., Schindler, S., Lencinas, M. V., Peri, P. L. and Martínez Pastur, G. 2016. Why biodiversity increases after variable retention harvesting: a meta-analysis for southern Patagonian forests. *For. Ecol. Manag.* 369(1), 161–9. doi:10.1016/j.foreco.2016.02.036.

Soliani, C. and Marchelli, P. 2017. *Zonas genéticas de lenga y ñire en Argentina y su aplicación en la conservación y manejo de los recursos forestales.* INTA Ed., Bariloche, Argentina.

Spagarino, C., Martínez Pastur, G. and Peri, P. 2001. Changes in *Nothofagus pumilio* forest biodiversity during the forest management cycle: 1. Insects. *Biodivers. Conserv.* 10(12), 2077–92. doi:10.1023/A:1013150005926.

Strassburg, B. B. N., Kelly, A., Balmford, A., Davies, R. G., Gibbs, H. K., Lovett, A., Miles, L., Orme, C. D. L., Price, J., Turner, R. K. and Rodrigues, A. S. L. 2010. Global congruence of carbon storage and biodiversity in terrestrial ecosystems. *Conserv. Lett.* 3(2), 98–105. doi:10.1111/j.1755-263X.2009.00092.x.

Sun, G. and Vose, J. 2016. Forest management challenges for sustaining water resources in the Anthropocene. *Forests* 7(12), 68. doi:10.3390/f7030068.

Swetnam, T. W., Allen, C. D. and Betancourt, J. L. 1999. Applied historical ecology: using the past to manage for the future. *Ecol. Appl.* 9(4), 1189–206. doi:10.1890/1051-07 61(1999)009[1189:AHEUTP]2.0.CO;2.

Thompson, I. D., Okabe, K., Tylianakis, J. M., Kumar, P., Brockerhoff, E. G., Schellhorn, N. A., Parrotta, J. A. and Nasi, R. 2011. Forest biodiversity and the delivery of ecosystem goods and services: translating science into policy. BioScience 61(12), 972–81. doi:10.1525/bio.2011.61.12.7.

Torres, A. D., Cellini, J. M., Lencinas, M. V., Barrera, M. D., Soler, R., Díaz-Delgado, R. and Martínez Pastur, G. J. 2015. Seed production and recruitment in primary and harvested *Nothofagus pumilio* forests: influence of regional climate and years after cuttings. *For. Syst.* 24(1), e-016. doi:10.5424/fs/2015241-06403.

Uhlenbrock Jansse, M. and Rodríguez, A. 2005. Evaluación de la productividad primaria neta arbórea potencial y la arquitectura vegetal para una mejor producción caprina en el departamento de Piura. *Zonas Áridas* 9(1), 161–77.

van Mantgem, P. J., Stephenson, N. L., Byrne, J. C., Daniels, L. D., Franklin, J. F., Fulé, P. Z., Harmon, M. E., Larson, A. J., Smith, J. M., Taylor, A. H. and Veblen, T. T. 2009. Widespread increase of tree mortality rates in the western United States. *Science* 323(5913), 521–4. doi:10.1126/science.1165000.

Veblen, T. T., Donoso, C., Kitzberger, T. and Robertus, A. J. 1996. Ecology of southern Chilean and Argentinian *Nothofagus* forests. In: Veblen, T. T., Hill, R. S. and Read, J. (Eds), *The Ecology and Biogeography of Nothofagus Forests*. Yale University Press, New Haven and London, pp. 293–353.

Villalba, R., Luckman, B., Boninsegna, J., D'Arrigo, R. D., Lara, A., Villanueva-Diaz, J., Masiokas, M., Argollo, J., Soliz, C., LeQuesne, C., Stahle, D. W., Roig, F., Aravena, J. C., Hughes, M. K., Wiles, G., Jacoby, G., Hartsough, P., Wilson, R. J. S., Watson, E., Cook, E. R., Cerano-Paredes, J., Therrell, M., Cleaveland, M., Morales, M. S., Graham, N. E., Moya, J., Pacajes, J., Massacchesi, G., Biondi, F., Urrutia, R. and Martinez Pastur, G. 2010. Dendroclimatology from regional to continental scales: understanding regional processes to reconstruct large-scale climatic variations across the western Americas. In: Hughes, M., Swetnam, T. and Díaz, H. (Eds), *Dendroclimatology: Progress and Prospects. Series: Developments in Paleoenvironmental Research*. Springer, Amsterdam, Holland, pp. 175–227.

Vilardy, S. P., González, J. A., Martín-López, B. and Montes, C. 2011. Relationships between hydrological regime and ecosystem services supply in a Caribbean coastal wetland: a social- ecological approach. *Hydrol. Sci. J.* 56(8), 1423–35. doi:10.1080/0262666 7.2011.631497.

Wallem, P. K., Anderson, C. B., Martínez Pastur, G. and Lencinas, M. V. 2010. Using assembly rules to measure the resilience of riparian plant communities to beaver invasion in subantarctic forests. *Biol. Invas.* 12(2), 325–35. doi:10.1007/s10530-009-9625-y.

Watson, J. E. M., Evans, T., Venter, O., Williams, B., Tulloch, A., Stewart, C., Thompson, I., Ray, J. C., Murray, K., Salazar, A., McAlpine, C., Potapov, P., Walston, J., Rosinson, J. G., Painter, M., Wilkie, D., Filardi, C., Laurance, W. F., Houghton, R. A., Maxwell, S., Grantham, H., Samper, C., Wang, S., Laestadius, L., Runting, R. K., Silva-Chávez, G. A., Ervin, J. and Lindernmayer, D. 2018. The exceptional value of intact forest ecosystems. *Natl. Ecol. Evolut.* 2(4), 599–610. doi:10.1038/s41559-018-0490-x.

World Resources Institute (WRI). 2003. *Earth Trends*. Washington Research Institute, Washington DC.

Whittingham, M. J. 2011. The future of agrienvironment schemes: biodiversity gains and ecosystem service delivery? *J. Appl. Ecol.* 48(3), 509–13. doi:10.1111/j.1365-2664.2011.01987.x.

Zhao, M. and Running, S. W. 2010. Drought-induced reduction in global terrestrial net primary production from 2000 through 2009. *Science* 329(5994), 940–3. doi:10.1126/science.1192666.

www.ingramcontent.com/pod-product-compliance
Lightning Source LLC
Chambersburg PA
CBHW050529270326
41926CB00015B/3142